Where on Earth!

A Gardener's Guide to Growers of Specialty Plants in California

agAccess
Davis, CA

Copyright © 1993 by **agAccess**

All rights reserved. No part of this book may be used or reproduced in any manner whatsoever without written permission from the publisher, except in the case of brief quotations embodied in articles or reviews.

Printed in the United States of America.

ISBN 0-932857-14-0

To order additional copies of this book, please call or write to:

agAccess
603 4th Street
Davis, CA 95616
(916) 756-7177
Fax: (916) 756-7188

CONTENTS

I. Introduction	iv
II. Growers of Specialty Plants	
North Coast	1
Mendocino, Humboldt, Lake	
North Bay	30
Sonoma, Marin, Napa	
San Francisco Peninsula	73
San Francisco, San Mateo, Santa Clara	
East Bay	96
Alameda, Contra Costa	
Sacramento Valley	113
Sacramento, Butte, Yolo, Shasta, Sutter	
Foothills	132
El Dorado, Nevada, Calaveras, Placer	
Mountains	157
Siskiyou, Plumas, Placer, Nevada	
Central Coast	163
Santa Cruz, Monterey, San Luis Obispo, San Benito, Santa Barbara	
San Joaquin Valley	204
San Joaquin, Fresno, Tulare	
Southern California	214
Riverside, San Diego, Los Angeles, Orange, San Bernardino, Ventura	
III. Other Sources of Plants	254
Oregon & Washington Growers	255
Seeds & Bulbs	259
Family Farm Guides	265
IV. Horticultural Information	268
Societies & Groups	269
Schools	271
Master Gardener Programs	274
V. Index	276

INTRODUCTION

Where on Earth! is a book for gardeners who believe that their gardens deserve better than Impatiens and Poison Oak, currently California's leading exotic and native ground covers. Six years of drought, renegade frosts and a robust herbivorous population have finally convinced everyone that California presents special challenges for gardeners. Fortunately we have a legion of entrepreneurial plant growers to meet these challenges. They seek, save and propagate plants suited to California's many micro-climates. They appreciate the unique qualities of California's flora and the suitability of mediterranean-climate plants for our frost-free areas. This book is about them.

We define growers as owners of enterprises which propagate or grow on at least half of their stock. Most grow nearly all their material. We realize that each California community is well served by a local garden center, many of which buy plants from these growers. The scope of this book did not permit their inclusion.

We believe that the best way to buy a plant is to see it. Growers listed in this book admit retail customers, though often only during limited times or with special restrictions, such as "must visit with a landscape professional". Please make sure you have the whole picture before you go. It is always a good idea to call first. Customers should not expect full retail services from wholesale businesses.

Each grower has a story to tell, and we hope we have told it accurately. They are as unique as the plants they offer. From diverse backgrounds, all are horticulturally well trained and educated. They all share a passion for what they grow. Their nurseries are generally small, family-run, and often result from a hobby that got delightfully out-of-control.

Growers are listed by geographic region to help you plan your shopping trip. A plant index in the back of the book will help you find what you are looking for. Horticultural attractions are listed at the end of each regional section. These public gardens, state parks, botanical gardens, and arboreta give you a chance to see mature specimens well-suited to the area. Specialty plant sales held by these groups are noted. Also, growers' recommendations of nearby places to visit are included in their listings.

Section III lists other sources of plant material: Oregon and Washington growers too good to exclude, even though they stretch our geographical limits; seed and bulb growers whose operations are strictly mail order; and guides to family farms and produce growers.

Section IV includes information about horticultural groups to join, horticultural programs offered by California schools and colleges, and Master Gardener programs.

We have visited most, but not all, of the nurseries listed. All have been recommended to us by respected professional horticulturists, but there are no guarantees that you will find each entry enchanting. Please let us know what you think. You are our greatest resource for future editions of *Where on Earth!*

This book is the result of nine years of research undertaken on behalf of the San Francisco Landscape Garden Show, a springtime horticultural spectacle. Proceeds from it, as well as from this book, benefit city parks in San Francisco. See you at the Show!

April, 1993   Nancy Conner
                                           Barbara Stevens
                                           Editors

CALIFORNIA REGIONS

This directory combines California's counties into regions. Each chapter in Section II begins with a more detailed map indicating growers' locations, referenced by page number.

- **A.** North Coast
- **B.** North Bay
- **C.** San Francisco Peninsula
- **D.** East Bay
- **E.** Sacramento Valley
- **F.** Foothills
- **G.** Mountains
- **H.** Central Coast
- **I.** San Joaquin Valley
- **J.** Southern California

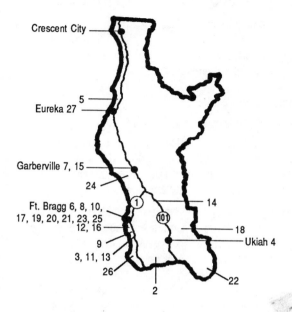

NORTH COAST

**MENDOCINO
HUMBOLDT
LAKE**

ANDERSON VALLEY NURSERY

18151 Mountain View Road
Boonville 95415
(707) 895-3853
Kenneth R. Montgomery, M.S.

Retail and Wholesale

Plant Specialties: California native plants (especially of the north coast), Mediterranean plants, including Cistus (over 40 species and hybrids), Rosemary, Lavandula, Helianthemum (20 cultivars), Daphne. Herbs, Perennials, and Drought-tolerant plants.

History/description: In 1978 Ken Montgomery gave up a career in research and teaching about plant ecology, figuring it was time to grow what California needs. The result is a splendid four acre farm in the beauty and tranquility of Mendocino County. This self-described "micro-business" aims to grow the highest quality plant material in an environmentally sound, low-tech way. It obviously works. Anderson Valley has introduced seven varieties of Cistus, Lavandula 'Lisa Marie', and is collaborating with UC to develop a variety of Olea europea suitable for the climate and soils of northern California. Ken recommends combining your visit with a stop at the gardens of the Boonville Hotel and local wineries.

How to get there: From Highway 101 in Cloverdale (about 30 miles north of Santa Rosa), take Highway 128 through the mountains about 30 miles northwest to Boonville. Continue through Boonville and make a left turn onto Mountain View Road. Nursery is on the left.

General information: Open 9 am - 5 pm; closed Wednesday and Sunday. Nursery will ship plants from the Bay Area to Eureka and from the Central Valley to the Coast. Slide talks available for interested groups.

County: Mendocino

DIGGING DOG NURSERY

P. O. Box 471,
Middle Ridge Road
Albion 95410
(707) 937-4389
Gary Ratway, Debra Wigham

Wholesale

Plant Specialties: Top select Perennials (especially late summer and fall bloomers), deciduous Asiatic woody plants, Heaths and Heathers (Erica, Calluna), Ceanothus, Arctostaphylos, Ornamental Grasses, Trees and Shrubs.

History/description: A founding member of Mendocino Coast Botanical Gardens, Gary Ratway acquired an interest in native and mediterranean plants from his days as a Landscape Architecture student at UC Davis. Originally next door to the Botanical Gardens, the nursery moved to its present, larger location in 1988 and is credited with the introduction of Oenothera 'Siskiyou' and Cistus 'John Patterson'. While in the area, in addition to Mendocino Coast Botanical Gardens, plan to visit Jug Handle Creek State Park.

How to get there: Call for directions

General information: Open to wholesale customers, by appointment only. Retail customers may visit with their landscaper. Plant lists are available for S.A.S.E. After 6/93, they will have a retail catalogue for mail order. They ship plants, and, if prompted, will share their horticultural knowledge with interested groups.

County: Mendocino

EVERGREEN GARDENWORKS

430 North Oak Street
Ukiah 95482
(707) 462-8909
Brent Walston, Susan Meier

Retail and Wholesale

Plant Specialties: Bonsai, especially Maples, Alpines and Rock Garden plants, deer and drought resistant plants, Mediterranean plants, Perennials, Ornamental Trees and Shrubs, including 25 varieties of Quince.

History/description: Started in 1987, this nursery was an outgrowth of a landscape maintenance, construction and design business. The unavailability of suitable, choice plants resulted in the decision to grow them on their own. Brent Walston continues to teach ornamental horticulture and bonsai culture at Mendocino community college and will give talks to interested groups. In addition to their display garden, they suggest you see the garden at Cleland House in Ukiah and Fetzer Vineyard garden Project in Hopland.

How to get there: From Highway 101 at Ukiah, take Perkins Street exit, go west one block past courthouse, right on Oak Street.

General information: Open from March 1 to Thanksgiving, 9 am - 3 pm. During winter months and for wholesale trade - by appointment only. They will ship plants. Plant list is available for $1.

County: Mendocino

FAIRYLAND BEGONIA AND LILY GARDEN

1100 Griffith Road
McKinleyville 95521
(707) 839-3034
Winkey Woodriff

Retail and Wholesale

Plant Specialties: Rhizomatous Begonias (15 varieties). Oriental, Asiatic and Aurelian Lilies (100 varieties)

History/description: Luther Burbank legacy lives. Leslie Woodriff started growing lilies, gloxinias and other bulbous plants after his childhood studies about this famous plantsman. He has been in the nursery business for 70 years, at this present location since 1982. His daughter Winkey bought the business in 1990, but he still actively hybridizes lilies. The freeze of 1989 hard hit their begonia plants, but they are rebuilding their stock. In 1991 Leslie Woodriff was presented with the Dix award, Holland's highest award for bulb hybridizing. Plan to visit the just beginning Humboldt Botanic Garden south of Eureka, the Azalea State Preserve on North Bank Road in McKinleyville, and the Carson Mansion in Eureka.

How to get there: On Highway 101 north after you have crossed Mad River Bridge, keep on Highway 101 by bearing left. Take School Road exit. Go right on Salmon, right on Griffith.

General information: Open hours are irregular. Call before you go. They will ship mail orders. Catalogue is available for $1.

County: Humboldt

FUCHSIARAMA

23201 North Highway One
Fort Bragg 95437
(707) 964-0429
Howard Berry

Retail

Plant Specialties: Fuchsias (40 species, 600 varieties). All plants rated for tolerance to heat, cold and dry conditions. Pelargoniums (60 varieties).

History/description: This retired meteorologist bought his wife one fuchsia plant in 1980 and became an overnight hobbyist. He has transferred his years of expertise as a weatherman to the scientific monitoring of the hardiness of fuchsia varieties. He bought a struggling fuchsia business in 1985 and has given it a new life - to say nothing of his own. He oversees a 13,000 square foot greenhouse and a 5-acre park-like setting where picnics, barbecues, horseshoes, and volleyball are encouraged. Should you develop a need for almost anything emblazoned with a fuchsia design (including gargoyles with sunglasses), you will find it in the gift shop.

How to get there: The nursery is located two miles north of Fort Bragg on Highway 1.

General information: Open Winter, everyday 9 am - 5 pm; Summer, everyday 8 am - 8 pm.

County: Mendocino

North Coast

GREENMANTLE NURSERY
3010 Ettersburg Road
Garberville 95542
(707) 986-7504
Marissa and Ram Fishman

Retail

Plant Specialties: 275 species of Old Roses, Iris, many other unusual landscape plants, Fruit Trees, including 130 varieties of apples (pink meat apples, sweet crabapples).

History/description: In the late 1800s this site was a major fruit breeding center. Its apples, the area's primary crop, fell victim to the coddling moth and competition from Washington state. Lately a new breed of orchardists are buying the old homesteads and reintroducing the interesting fruit developed by Albert Etter, pioneer fruit breeder, often described as the Luther Burbank of Humboldt County. The Fishmans are involved in the task of retrieving old Etter varieties. Marissa started their collections of roses and iris. Primarily a catalogue business since 1983, Greenmantle permits visitors to come in late spring to see the display garden on this three acre site.

How to get there: Call for directions.

General information: Open only by appointment in late May and early June. Greenmantle ships plants and has a rose catalogue available for $3, and a plant list of fruit trees for S.A.S.E. They happily speak on their subject.

County: Humboldt

HEARTWOOD NURSERY

525 South Franklin
Fort Bragg 95437
(707) 964-3555
Dan Charvet, Patty Leahy

Retail

Plant Specialties: Perennials, including hardy Geraniums, Lavandulas. Heaths and Heathers (25 varieties). Dwarf conifers, including Chamaecyparis, Pinus, Cryptomeria. Japanese Maples, Camellias, Ornamental Grasses, Anemones, Abutilons.

History/description: After college both partners found work in nurseries, eventually becoming certified California nurseryworkers. They bought a pre-existing nursery business in 1983 and have spent the time since experimenting with unusual and rare seed. They will also help you plan your garden and recommend a visit to the Mendocino Coast Botanical Gardens and Jughandle Creek State Park.

How to get there: From Highway 101 North in Fort Bragg, go right on Chestnut Street. After one block, go right on Franklin Street.

General information: Open Monday through Saturday, 10 am - 5 pm . During March through August, open Sunday, noon - 4 pm. Plant list is available for S.A.S.E.

County: Mendocino

HERITAGE HOUSE NURSERY

5200 North Highway 1
Little River 95456
(707) 937-1427
Peggy Quaid, Manager

Retail. Some wholesale to locals only.

Plant Specialties: Old-fashioned, rare and unusual Perennials. Woodland California native plants, Ground Covers, Grasses, Mediterranean shrubs, Heaths and Heathers, Rock Garden plants.

History/description: Since 1985 visitors to Heritage House have had the opportunity to buy plants similar to those used to landscape the 37 acre resort. The garden is laid out along a series of trails which connect forested areas and woodland gardens to the coastal bluff with its heather and rock gardens. They recommend another magical place, down Highway 1 in Manchester - Jim Thompson's private heather garden. (707-882-2345, call for appointment, best visiting times are May, August or September).

How to get there: Take Highway 101 north to Highway 128 to Highway 1. Go north on Highway 1 for 5 or 6 miles.

General information: Open everyday, 10 am - 3 pm.

County: Mendocino

HERITAGE ROSE GARDEN

16831 Mitchell Creek Drive
Fort Bragg 95437
(707) 964-3748
Joyce and Gary Demits,
Virginia and Howard Hopper

Retail

Plant Specialties: Well over 150 varieties of Antique Roses, grown on their own roots. They maintain an index of roses which are virus-free for six years. All field-grown, sold 2 - 3 years old. Specializing in Victorian Ramblers and the re-establishment of "lost" roses.

History/description: Operating primarily a mail-order business, these sisters do some selling on-site. They have always grown roses as a hobby, attracted at first to antique ones because they required less care than their modern counterparts. Membership in the Heritage Rose Group encouraged their first business effort in 1981. They have become increasingly interested in the romance and history of old roses, while working on producing an index of virus free roses. These two indefatigable experts take care of all operations all by themselves to insure quality. They maintain a display garden with over 1,000 varieties. They recommend you take the walking tour of Mendocino during rose season. They will supply details.

How to get there: Be sure to call first. From Highway 1, 1/4 mile north of the Botanical Garden, go right on Simpson Lane. Right on Mitchell Creek after 2.2 miles. They are the fourth house on the left. Pull rings of deer gate toward you.

General information: Open by appointment only. Catalogue available for $1.50. They will ship mail orders.

County: Mendocino

LORRAINE'S GERANIUM COLLECTION & WOODS FARM NURSERY

P. O. Box 311
Albion 95410
(707) 937-4066
Lorraine and David Woods

Retail and Wholesale

Plant Specialties: Over 300 varieties of Pelargoniums - dwarf, fancy leaf, rosebud, cactus-flowered - and 20 varieties of scented pelargoniums.

History/description: In 1983 Lorraine and David Woods retired from teaching and working in the public schools. Both were interested in plants and so began a conifer nursery as an income-producing retirement venture. As their interests switched so did their plant stock. They currently grow all of their pelargoniums from cuttings.

How to get there: Call for directions.

General information: Open by appointment only Plant list is available for S.A.S.E.

County: Mendocino

MENDOCINO HEIRLOOM ROSES

Box 670
Mendocino 95460
(707) 937-0963
Alice Flores, Gail Daly

Wholesale & Retail

Plant Specialties: Heirloom roses found or raised before 1900. An especially fine collection of ramblers.

History/description: Rosarian Alice has long been fascinated by the old roses growing in and around the town of Mendocino. She and friend Gail determined to preserve and disseminate these roses planted by the area's settlers. They now propagate many other roses as well. Alice and Gail emphasize that old roses lend themselves very well to organic gardening methods, which they employ at their farm in Redwood Valley. You can also request their *Walking Tour of Mendocino Roses* ($1).

General information: Mail order only. They ship bare-root between December 30 and March 1. They will also deliver wholesale orders in pots to nurseries in Northern California. Their very informative catalogue is available for $1.

County: Mendocino

MOON RIVER NURSERY

31901 Middle Ridge Road
Albion 95410
(707) 937-1314
Sharon Hanson, Tom Wodetsky, Jess River

Wholesale

Plant Specialties: Perennials (Campanula, hardy Geranium, Penstemon, Salvia, Hellebore), Mediterranean plants (Cistus, Lavandula, Rosemary). Trees, Vines, and Ground Covers.

History/description: Just open since 1988, Moon River's partnership combines horticultural knowledge and business savvy with managerial skills. Sharon Hanson had previously been working at another local nursery for ten years so well knows what grows on the North Coast. Planted out "mother beds" enable visitors to see mature specimen plants.

How to get there: Call for directions.

General information: Open by appointment only, Monday through Friday, 9:30 am - 5 pm. They will deliver to the Bay Area every other week for minimum orders of $250. Plant list available for S.A.S.E.

County: Mendocino

MOUNTAIN MAPLES

P. O. Box 1329
Laytonville 95454-1329
(707) 984-6522
Nancy and Don Fiers

Retail and Wholesale

Plant Specialties: Japanese Maples (1-, 2-, 3-year old bench grafted Acer palmatum for bonsai and for landscape use), some Acer buergeranum, Acer circinatum 'Little Gem'.

History/description: Owning a beautiful property and wanting to work at home, the Fiers decided to go into the nursery business. In no rush, they resolved to specialize in slow growth plants. Then they met J. D. Vertrees, author of a well known book on the genus Acer. Under his aegis, maples became their exclusive focus. They now have 20,000 plants in pots with another 8,000 in cold frames. They recommend a visit to Richardson Grove State Park and the Nature Conservancy walk on Branscomb Road toward the coast from Laytonville.

How to get there: They are on a dirt road 10 miles north of Laytonville. Call for directions.

General information: Open by appointment only. They will ship plants, have a catalogue available for $1, and will give talks to interested groups.

County: Mendocino

NANCY R. WILSON SPECIES AND MINIATURE NARCISSUS

6525 Briceland-Thorn Road
Garberville 95542
(707) 923-2407
Nancy R. Wilson

Retail

Plant Specialties: Unusual species of narcissus, all grown from seed: none wild-collected. Special attention to hybrids of special colors, smaller sizes, and Rock Garden-compatible plants.

History/description: When you grow up with a mother who has a well-developed bulb habit and the first house you buy with your hasband just happens to have a rock garden packed with bubls, you are destined to be a bulb lover. This health care professional and her retired husband succumbed to their destiny and started a nursery in 1990. Nancy Wilson recommends you visit and support the embryonic Humboldt County Botanical Garden at the College of the Redwoods.

How to get there: Call for directions.

General information: Open by appointment anytime although best bloom season is March and April. Mail order catalogue available for $1, deductible from purchase. Site-specific consultation by mail only.

County: Humboldt

NATIVE UPRISINGS

32033 Middle Ridge Road, or P. O. Box 1174
Mendocino 95460
(707) 973-3903
Alison Gardner, Paul Kish

Retail and Wholesale

Plant Specialties: California native plants, especially those suited to north coast.

History/description: These partners with their backgrounds in taxonomy, botany and native plants started this nursery in 1992. Previously working at Moon River Nursery, they decided to strike out on their own and specialize in natives only. Large stock plants, from which they propagate mostly by division and cuttings, substitute for an official display garden. They will custom grow and offer site-specific consultations.

How to get there: Call for directions.

General information: Open everyday, by appointment only. Catalogue is available for $2 and they will do limited shipping. They will be pleased to give a talk to your group.

County: Mendocino

NOYO HILL NURSERY

31720 Highway 20
Fort Bragg 95437
(707) 964-9308
Steve and Nicole Sloan

Retail and Wholesale

Plant Specialties: Rhododendron (200 species), Azaleas, Heathers (20 species), Ferns.

History/description: The Sloans bought this nursery in 1989, a logical tie-in to their landscaping business and entrepreneurial bent. In business as Lauer's Nursery since 1935, the nursery now has 30,000 mainly ericaceous plants covering 2.5 acres in various sizes up to 25 gallon. Plan to visit Hare Creek Nature Trail in Jackson State Forest when you are in the area.

How to get there: From the intersection of Highway 1 and Highway 20 south of Fort Bragg, travel 1.2 miles east on Highway 20.

General information: Open everyday, 9 am - 5 pm.

County: Mendocino

RARE CONIFER NURSERY

P. O. Box 100, Dept. H.
Potter Valley 95469
John Mayginnes

Retail and Wholesale

Plant Specialties: Rare, endangered and threatened Conifer species, Larix, Abies, Picea, ungrafted whenever possible - seeds and seedlings to 30" boxes.

History/description: Resourceful businessman John Mayginnes has combined his love of conifers with an investment banking career on and off Wall Street. Relocated to Ukiah in 1972 to start a chemical company, he bought a piece of land nearby in the remote mountainous area near the Mendocino National Forest and founded the nursery. Since 1989 he has been hard at work developing an 80 acre arboretum of conifer species, called Garden of Kirpal (meaning "compassion" in Hindi). He is also founder and director of the Rare Conifer Foundation, a non-profit, global seed bank, dedicated to protecting conifers from extinction in their native habitats. Funds raised help defray the cost of worldwide, plant collecting expeditions.

How to get there: Write for appointment. This remote nursery has no telephone, but please give your phone number for return call.

General information: Open by appointment only. They can ship any order. Catalogue available upon request.

County: Mendocino

REGINE'S FUCHSIA GARDENS AND THE ORCHID BENCH

32531 Rhoda Lane
Fort Bragg 95437
(707) 964-0183
Regine and Bruce Plows

Retail

Plant Specialties: 300 Fuchsia varieties, hard-to-find species, such as F. exchorticata, F. venusta, F. procumbens, F. magellanica var. molinae. Old and new hybrids, including 'Purple Rain', 'Pink Rain', 'Venus Victrix', 'Lye's Unique'. The Orchid Bench specializes in Orchid seedlings and mericlones of a number of genera, emphasizing Paphiopedilum.

History/description: These two orchid hobbyists left the Los Angeles area in 1980 for Fort Bragg. Well-known fuchsia hybridizer Annabelle Stubbs' nursery and home had just gone on the market. They fell in love with it and bought it which explains this combination of plant specialties. This small operation with small display garden stresses quality over quantity. Plan also to visit the Ecological Staircase and Pygmy Forest in Jughandle State Reserve.

How to get there: From Highway 1 south of Fort Bragg, go east on Simpson Lane, south on George's Street and west on Rhoda.

General information: Open May to October, Wednesday through Sunday, 10 am - 5 pm. Call ahead in winter. Mail order catalogue for fuchsia cuttings is available for $1. Cuttings are made to order from January to June; allow six weeks.

County: Mendocino

RICHARDS LANDSCAPE AND GARDENING

32101 Highway 20
Fort Bragg 95437
(707) 964-0710
Charles Richards

Retail and Wholesale

Plant Specialties: Rhododendrons, exclusively Exbury seedlings, varieties and some named cultivars, plus R. occidentale in a wide variety of colors.

History/description: This family business was started by Charles Richards' father who began hybridizing rhododendrons and azaleas as a hobby in 1952. He is still at it although the business has been run by the son since 1968. March through May is the optimum time to visit for peak blossoms and the powerful fragrance of California's native azalea, R. occidentale. They will custom grow for clients.

How to get there: The nursery is on Highway 20, which connects Highway 1 with Highway 101 between Fort Bragg and Willits.

General information: Open everyday, 8 am - 5 pm.

County: Mendocino

SHERWOOD NURSERY

30480 Sherwood Road
Fort Bragg 95437
(707) 964-0800
Mike Peterson

Retail and Wholesale

Plant Specialties: Rhododendrons, Ponticum hybrids, Maddenii, Yakushimanum crosses.

History/description: Trained as a forester, Mike Peterson ran a Georgia Pacific tree nursery for 17 years. During this time he was a rhododendron hobbyist and when the itch got to be "too much", he opened this primraily wholesale business in 1984. Public visitors are welcome by appointment only and should not expect full retail services. He is the current president of the local Rhododendron Society.

How to get there: Call for directions.

General information: Open by appointment only. Plant list is available.

County: Mendocino

SPECIALTY OAKS

12552 Highway 29
Lower Lake 95457
(707) 995-2275
John McCarthy

Wholesale

Plant Specialties: Native oaks (Quercus, 5 species) Each year 2,500 choice examples are selected from approximately 9,000 seedlings. Shaped and field-grown for 5 years, the result is a well-proportioned 8' - 9' oak with a 2" caliper.

History/description: Involved with a peninsula tree care service, John McCarthy noticed the difficulty of replacing the many oaks falling victim to age and suburban encroachment. Seizing the opportunity, he switched to nursery production in 1985, although he still gives tree care consultations.

How to get there: Call for directions.

General information: Open to wholesale customers, by appointment only.

County: Lake

SUMMERS LANE NURSERY

20000 Summers Lane
Fort Bragg 95437
(707) 964-7526
James and Don Celeri

Retail and Wholesale

Plant Specialties: Rhododendrons, 100 varieties, species, hybrids. including R. macrophyllum and R. occidentale (California native rhododendrons) and R.'Noyo Dream', their cross between R. yakushimanum and R. 'Mars'.

History/description: These brothers both worked in the lumber industry and realized there was need for a career change but did not want to foresake their relationship with plants. Local enthusiasm for rhododendrons and their suitability to the area convinced them to grow this plant. They have been in business since 1989 and recommend you schedule your visit to coincide with the Rhododendron Show held in the last week in April at the Dana Gray Elementary School in Fort Bragg. With 700 entries, it is the largest show on the West Coast.

How to get there: Call for directions.

General information: Open everyday, by appointment only. Plant list available for S.A.S.E.

County: Mendocino

TIERRA MADRE

545 Shelter Cove Road
Whitethorn 95489
(707) 986-7215
Jayme Stark

Retail

Plant Specialties: California native plants, especially those suited to northern California coastal and inland areas. Drought tolerant plants, Perennials (200 varieties), Annuals, Vegetables, all organically grown.

History/description: Jayme Stark had sixteen years of experience in landscaping and owned a lumber business which carried gardening supplies, when customers suggested she open a nursery. She acquiesced to their wishes in 1992.

How to get there: Take Highway 101 north past Garberville to Redway exit. Take Shelter Cove/Briceland Road. Go west 15 miles to Shelter Cove/Whitethorn junction. Continue half mile more to nursery.

General information: Open Tuesday through Saturday, 10 am - 4:30 pm, March through November.

County: Humboldt

TRILLIUM LANE

18855 Trillium Lane
Fort Bragg 95437
(707) 964-3282
Eleanor Philp

Retail

Plant Specialties: Rhododendrons, species, varieties and hybrids. Some companion plants.

History/description: A doyenne of rhododendron culture, Eleanor Philp started this nursery in 1962 as an outgrowth of her plant collecting hobby. She is a preeminent grower and hybidizer and is credited with the naming of R. 'Fort Bragg Centennial' and R. 'Fort Bragg Glow'. There are over 600 varieties in her display garden. Now slowing down just a bit, she has curtailed her hybridizing efforts but not her enthusiasm or her love of subject. She gives talks and tours of her domain.

How to get there: Call for directions.

General information: Open by appointment only.

County: Mendocino

WATCHWOOD GARDEN DESIGN & NURSERY

P.O. Box 21
Pt. Arena 95468
(707) 882-2415
Marilyn Bucher

Retail and Wholesale (orchids - retail only)

Plant Specialties: Orchids, rare plants, and Bulbs.

History/description: Marilyn Bucher is a delightful, self-proclaimed plant addict who has been selling the labors of her love since 1981. Watchwood is located in the wild woods of the south Mendocino coast overlooking the sea.

How to get there: Call for directions.

General information: Open by appointment only, generally Wednesday and Saturday, 9 am - 5 pm. Plants delivered from Point Arena to Bay Area. Future plans include a catalogue and shipping.

County: Mendocino

WESTGATE GARDEN NURSERY

751 Westgate Drive
Eureka 95503
(707) 442-1239
Catherine Weeks

Retail and Wholesale

Plant Specialties: Rhododendrons and Azaleas, species and hybrids, Chionanthus, Franklinia, Halesia, Styrax, Stewartia, Maples, Disanthus, Drimys winteri, Desfontainea, Eucryphia, Embothrium coccineum, Enkianthus 'Red Bells', Exochorda, Fothergilla, Carpenteria californica 'Elizabeth', Cornus capitata. Michelia, Leucothoe, Garrya, Kalmia, Viburnum.

History/description: Catherine Weeks continues the business she started with her husband in 1963. Their first home had a greenhouse and lath house which encouraged their interest in plants. Intensely curious and largely self-taught, they became successful cutting propagators. This huge collection of trees and shrubs fits on just a 2.5 acre site. Their stock plants combine to create a display garden. She recommends you also see the native azaleas when in bloom on Stagecoach Hill.

How to get there: From Highway 101 south of Eureka, take Elk River exit east. In 50' go right on Elk River Road. After 1.5 miles, go left on Westgate Drive.

General information: Open everyday except Wednesday, 8:30 am - 6 pm. They will ship plants. Catalogue available for $4, refundable with purchase.

County: Humboldt

HORTICULTURAL ATTRACTIONS

Mendocino Coast Botanical Gardens

18220 North Highway One
Fort Bragg 95437
(707) 964-4352

History/description: Originally the private garden of Ernest and Betty Schoefer, these 47 landscaped acres are now part of the Mendocino Coast Recreation and Park Department, but supported entirely by memberships, entrance fees and volunteer help. Located in the moist, coastal pine forest, their specialty is Rhododendrons. Other noteworthy plant collections include heaths and heathers, ivies, and perennials. The topography is varied, ranging from breath-taking coastal bluffs to fern-filled canyon trails. They offer garden walks, evening lectures, two-hour docent tours for groups, and, of course, the opportunity to volunteer. Adult admission is $5 which includes a self-guided walking tour.

General information: Open everyday but Thanksgiving and Christmas. November through February, 9 am - 4 pm; March through October 9 am - 5 pm.

Plant sales: Garden store sells plants from their nursery. Big plant sale held twice a year.

County: Mendocino

Other Attractions:

Fetzer Vineyards Garden Project at Valley Oaks

13601 Eastside Road
Hopland 95449
(707) 744-1250

California adaptation of European-style garden; vegetable, flower, fruit and herb gardens.

Kruse Rhododendron State Reserve

20 miles north of Jenner, 10 miles south of Gualala
(707) 865-2391

Native rhododendrons colonizing second growth forest.

Jughandle State Reserve

Three miles north of Mendocino on Highway One
(707) 937-5804

Ecological staircase, self-guided nature trail from beach to pygmy forest.

Montgomery Woods State Reserve

Five miles east of Comptche on Comptche-Ukiah Road
(707) 937-5804

Huge coast redwoods.

■30 **Where On Earth!**

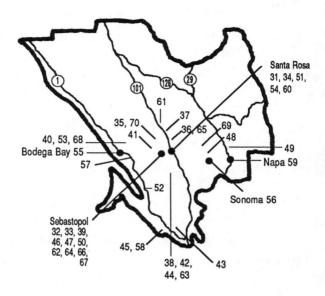

NORTH BAY

**NAPA
SONOMA
MARIN**

ALPINE VALLEY GARDENS

2627 Calistoga Road
Santa Rosa 95404
(707) 539-1749
Dorothy and Wilbur Sloat

Retail

Plant Specialties: Daylilies

History/description: All the daylilies here are field grown - dug up while you wait and sold bare root with 2 or more fans per plant. The American Hemerocallis Society's Display Garden at the nursery's entrance has over 300 varieties, including all the winners of the Society's yearly, best daylily award, the Stout Medal. When they finally breed that pure white or blue daylily, you will see it here first.

How to get there: Highway 101 to Highway 12 exit. Take Highway 12 to Calistoga Road. Take Calistoga Road toward Calistoga for 4 miles.

General information: Open anytime, by appointment only. The Display Garden is open from June 15 - July 15, daily from noon to 5 pm. A catalogue of plants is available for S.A.S.E.; they will ship anywhere. Also, they can give slide talks to interested groups.

County: Sonoma

APPLETON FORESTRY

1369 Tilton Road
Sebastopol 95472
(707) 823-3776
Harold and Patricia Appleton

Retail and Wholesale

Plant Specialties: California native plants, mostly Trees, some Shrubs for shade.

History/description: A registered state forester and erosion control expert, Harold Appleton was well placed to notice the unavailability of native trees. He still produces revegetation and mitigation plans, while Patricia manages the nursery. Together they propagate all their plants in super-cels, dee-pots, and tree-pots and are available for site-specific consultations.

How to get there: Call for directions.

General information: Open by appointment only. Call in early afternoon.

County: Sonoma

BAMBOO SOURCERY

666 Wagnon Road
Sebastapol 95472
(707) 823-5866, FAX (707) 829-8106
Gerald Bol

Retail

Plant Specialties: Bamboo

History/description: Like many specialty nurseries, the origins of Bamboo Sourcery lie in the plant collecting passion of one exceptional individual. Officially beginning in 1984, Gerald Bol has landscaped five sloping and terraced acres to display more than 300 forms and species of bamboo. Deservedly so, Gerald is President of the American Bamboo Society.

How to get there: From areas south of Sonoma County, take 101 north to first Sebastopol exit, Highway 116 west. At first traffic light in Sebastopol, turn left on Highway 12 going towards Bodega. Drive 4 miles and turn left onto Wagnon Road. Bamboo Sourcery is half mile on the left.

General information: Open by appointment only. Will ship anywhere. Catalogue available for $2. Plant list for S.A.S.E. Has slide talk for interested groups.

County: Sonoma

BURGANDY HILL NURSERY

4577 Hall Road
Santa Rosa 95401
(707) 579-3790
Becky and Kim Hill

Wholesale

Plant Specialties: Dwarf conifers, Ornamental grasses, California native plants, Daylilies, Shrubs, Trees and Perennials.

History/description: About 1986, speech therapist Kim Hill decided he would rather talk to plants than to people and started growing choice specimens at home in Sebastopol. Burgandy Hill Nursery now occupies ten acres west of Santa Rosa.

How to get there: Take Highway 101 to Highway 12 west toward Sebastopol. Turn right onto Fulton Road and go to second signal. Turn left onto Hall Road. The nursery is .8 miles on the right.

General information: Open to wholesale customers only. Public may visit with a landscaper. Monday through Friday, 8 am - 5 pm and Saturday by appointment. They deliver throughout Sonoma, Napa and Marin counties, San Francisco, the Peninsula, and to Vallejo, Berkeley and Walnut Creek. Current plant availability list sent upon request.

County: Sonoma

CALIFORNIA CARNIVORES

7020 Trenton - Healdsburg Road
Forestville 95436
(707) 838-1630
Peter D'Amato, Marilee Maertz

Retail and Wholesale

Plant Specialties: Over 400 varieties of Carnivorous plants, specializing in mature selection of pitcher plants, sundews, flytraps and butterworts.

History/description: Grower Peter D'Amato began his interest in insect-eating plants at age twelve growing up near the New Jersey Pine Barrens. The nursery was opened after his successful display at the 1989 San Francisco Landscape Garden Show. California Carnivores is the only nursery of its type open to the public and houses one of the world's large collections. There are picnic grounds in the beautiful setting of Mark West vineyard. Books, T-shirts and supplies are also for sale and garden groups may arrange tours.

How to get there: Highway 101 north past Santa Rosa to River Road exit. Turn left on River Road for 5.5 miles and take a right onto Trenton/Healdsburg Road at the Mark West Winery sign. Nursery located at Mark West Winery.

General information: Open everyday, 10 am - 4 pm. Call ahead in winter. They ship mail order plants.

County: Sonoma

CALIFORNIA FLORA NURSERY

2990 Somers Street,
P. O. Box 3
Fulton 95439
(707) 528-8813
Phillip Van Soelen, Sherrie Althouse

Retail and Wholesale

Plant Specialties: California native plants, unusual Perennials, Mediterranean plants.

History/description: These two individuals are committed to making native plants available to the gardening public. Sherrie is past president of her area chapter of the CNPS. Phil's background in environmental studies produced a child's ecology book, *"Cricket in the Grass"*, as well as a passion for native plants. Their fruitful partnership has been in business since 1981.

How to get there: Highway 101 north to River Road exit (first exit north of Santa Rosa). Go west on River Road for one mile. Go left on Somers Street just before the traffic light. Nursery at corner of Somers and D Street.

General information: Open Monday to Friday, 9 am-5 pm; Saturday, 10 am - 4 pm. Winter hours: Monday - Friday, 10 am - 4 pm. They will deliver wholesale purchases to the Bay Area and they have a free plant availability list.

County: Sonoma

CIRCUIT RIDER PRODUCTIONS, INC.

9619 Old Redwood Highway
Windsor 95492
(707) 838-6641, FAX (707) 838-4503
Betty Young, Manager

Retail and Wholesale

Plant Specialties: Oaks and other California native woody plants. Site-specific collection for restoration of disturbed or developed areas. Most sold in super cell tubes to encourage proper root development.

History/description: This non-profit organization has a win-win prospectus. They are revegetation specialists who for 15 years have propagated and sold plants to restore disturbed areas. Profits from this enterprise go to address human resource needs, such as job training, employment counselling, drug prevention, and summer youth employment. Not many enterprises set out to solve both the problems of the earth and its inhabitants. We laud their efforts and success.

How to get there: Take Highway 101 north past Santa Rosa to Windsor exit. Continue west on Old Redwood Highway for .6 miles.

General information: Open by appointment only.

County: Sonoma

COTTAGE GARDEN GROWERS
4049 Petaluma Boulevard North
Petaluma 94952
(707) 778-8025
Bruce Shanks, Mark Williams

Retail and Wholesale

Plant Specialties: More than 300 varieties of Perennials.

History/description: These lifelong friends with complimentary talents (Bruce's background is in horticulture, Mark's is in marketing) opened Cottage Garden Growers in 1990. Their lovely 3/4 acre property sits on a fieldstone-terraced hillside. The nursery grows on and propagates only selected perennials from around the world.

How to get there: From Highway 101 north, take Penngrove Exit. Petaluma Boulevard is in half mile.

General information: Open everyday, 9 am - 6 pm. Winter hours, 9 am - 5 pm. Write or call for their plant list.

County: Sonoma

EMERISA NURSERY

3150 A Mueller Road
Sebastopol 95472
(707) 823-5102
Muchtar Salzman

Retail and Wholesale

Plant Specialties: 500 species of Perennials in 4" pots, featuring Penstemon, hardy Geraniums, Cistus, Hebe. Also Herbs, Drought tolerant plants, Cold hardy plants, and some 1 gallon Shrubs.

History/description: After raising 8 children, working as a flagman, pizzamaker, and teacher, Muchtar Salzman could still muster the energy to return to school - Cal Poly - and earn a degree in crop science. After working as a field assistant for agricultural extension programs, he bought the two acre, former Fred Hall Nursery in 1990. The nursery is enriched by his lifetime of experiences gathered from growing up in Holland, and living in New Guinea and the eastern US. Truly a family man, he named the nursery for his parents - Emerson and Marisa - and has four of his brood and his brother working with him.

How to get there: Call for directions.

General information: Open Monday through Saturday, 7:30 am - 5 pm; Sunday, 10 am - 3 pm. Plant list is available for S.A.S.E.

County: Sonoma

ENJOY RHODODENDRONS

1890 Joy Ridge Road
Occidental 95465
(707) 874-3055
Paul Molinari

Wholesale

Plant Specialties: Rhododendrons

History/description: Located on a coastal woodland ridge, this well-tended nursery offers visitors the chance to see over 45,000 rhododendrons in perfect growing conditions. Known especially for their exceptional collection of maddeniis which they continue to select and evaluate, Enjoy has a wide selection of almost all the sections of the genus Rhododendron. Paul Molinari's initial interest in growing our native Rhododendron macrophyllum has gone global. This botanical garden of rhododendrons from all over the world was begun in 1975 in partnership with Doug Schaeffer.

How to get there: Call for directions.

General information: Open to wholesale customers, by appointment only. Public may visit with a landscape professional. Tours of the garden and nursery may be arranged in advance.

County: Sonoma

FARWELL AND SONS NURSERY

12983 Bodega Avenue
Freestone 95472
(707) 823-8415
Gary Farwell

Retail

Plant Specialties: Rhododendrons, hybrids selected for general west coast landscape use. R. 'Inheritance' hybridized by owner.

History/description: Conscripted for nursery chores since he was a child by his father Everett Farwell, well-known rhododendron grower in Woodside, Gary Farwell obviously thrived under forced labor. He has been in business on his own since 1977 and has several hybirds to his credit. All his plants are container-grown, in 5 and 7 gallon sizes. He recommends you visit the Kruse Rhododendron State Reserve near Fort Ross.

How to get there: From Highway 101, go west on Highway 12 which becomes Bodega Avenue. Continue through Sebastopol. Nursery is seven miles west of Sebastopol on the left side.

General information: Open Friday through Tuesday, 9 am - 5 pm.

County: Sonoma

GARDEN VALLEY RANCH NURSERY

498 Pepper Road
Petaluma 94952
(707) 795-5266
Ray Reddell, Rick Weeks

Retail and Wholesale

Plant Specialties: Roses and companion plants.

History/description: Roses were just a hobby of Ray Reddell's until 1981 when he began in earnest to plant fields of roses for the cut flower market. The nursery is a later addition, a wonderful complement to the cut flower business and a much-anticipated boon to Garden Valley's loyal customers. They now also have test gardens and a heritage rose garden. They offer self-guided tours for $3.

How to get there: Highway 101 north to Old Redwood Highway/Penngrove Exit. Go left back over freeway to first stop light. Go right on Stony Point for 1.5 miles. Go left on Pepper Road. Ranch on right.

General information: Open Wednesday through Saturday 10 am - 4 pm. Sunday 11 am - 4 pm. They have a mail order catalogue and will ship bare-root roses by mail. They give interesting talks to groups.

County: Sonoma

GERANIACEAE

122 Hillcrest Avenue
Kentfield 94904
(415) 461-4168
Robin Parer

Retail

Plant Specialties: Geraniums, Erodiums and other members of the geranium family.

History/description: Since 1984 this delightful "Geranium Lady" has passionately promoted the hardy geranium as a garden-worthy plant. She has been on plant-collecting trips to remote areas of Australia, Europe and the Andes. Her half acre woodland garden adjacent to the nursery gives visitors the opportunity to see 165 species and varieties of herbaceous geraniums and erodiums in a garden setting.

How to get there: Call for directions.

General information: Open by appointment only. Catalogue available for $2.50. Slide presentation available for interested groups.

County: Marin

THE GREAT PETALUMA DESERT
5010 Bodega Avenue
Petaluma 94952
(707) 778-8278
Jerry Wright

Retail and Wholesale

Plant Specialties: Cacti and Succulents, rare and unusual specimens, including x Pachgerocereus orcuttii (extinct in its habitat). Extensive Cycad collection. Also specializing in plants from Madagascar, Pachypodium, Beaucarnea, Euphorbia, Aloe, Haworthia. Up to 55 gallon sizes.

History/description: Courtesy of the U.S. Army, Jerry Wright moved to California in 1974 and got interested in cacti. As his personal collection and his contacts with fellow hobbyists grew, he started to help out at a local succulent nursery. He got a business license and tested the waters with a few plant orders. Response was so great that five years later he retired and entered the nursery business full-time. He maintains a display garden and can help you plan your garden.

How to get there: From Highway 101, take E. Washington exit. E. Washington Steet becomes Bodega Avenue as you leave central Petaluma. Nursery is approximately 5.5 miles from Highway 101.

General information: Open Friday, Saturday and Sunday, 10 am - 5 pm.

County: Sonoma

LARNER SEEDS

P. O. Box 407
Bolinas 94924
(415) 868-9407
Judith Lowry

Retail and Wholesale

Plant Specialties: California native plants, especially suited to the coast, Native Grasses, Seeds.

History/description: In business as a seed collector and source since 1977, Judith Lowry finally succumbed to the requests of landscapers for actual plants grown from her wide variety of native seeds. In the nursery business since 1986, she continues to perpetuate the horticultural gene pool for the revegetation of West Marin and the North Coast. She will custom grow plants from seed collected by landowners for site-specific restoration. She recommends that you allow time to visit the Audubon Canyon Ranch and Point Reyes Bird Observatory's self-guided nature trail.

How to get there: Call for directions.

General information: Open by appointment only. Mail order seed catalogue is available free of charge. Talks gladly given to interested groups.

County: Marin

LONE PINE GARDENS

6450 Lone Pine Road
Sebastopol 95472
(707) 823-5024
Ian C. Price, Janet M. Price

Retail and Wholesale

Plant Specialties: Cacti, Succulents and Bonsai, seedlings and cuttings of bonsai starters and field-grown, larger pre-bonsai stock.

History/description: The British background of this family mandated its gardening interest. Educated in botany and horticulture in the U.S. (specializing in desert plants), the Prices returned to England to operate a nursery there. A nostalgia for the opportunity to grow desert plants out-of-doors brought them back to the West. Their nursery opened in 1975 and affords visitors a spectacular panoramic view of the coastal mountains.

How to get there: Highway 101 north to Highway 116 west (Rohnert Park/Sebastopol Exit). Go 4.5 miles. Left on Lone Pine Road. Nursery is one mile at the top of the hill.

General information: Open Thursday, Friday, Saturday 10 am - 5 pm. They will deliver wholesale orders to the San Francisco Bay Area, Sacramento valley, foothills and Monterey area. A mail-order catalogue, "Lone Pine Connection," is available for $3 from P. O. Box 1338, Forestville 95436.

County: Sonoma

LONGDEN NURSERY

6179 Lone Pine Road
Sebastopol 95472
(707) 823-4535
Stewart Campbell, Steve Campbell

Wholesale

Plant Specialties: Camellias, South African and Australian plants, Perennials.

History/description: Longden Nursery can rightfully claim to have been in business for generations. Stewart Campbell's father started it in 1947 and now his son, Steve, is manager. Named for its original location on Longden Avenue in San Gabriel, the nursery moved to its present site in 1970. Known originally for its camellias, Longden Nursery now has a wide range of unusual, large specimen plants. Always on the lookout for new plant introductions from around the world, the Campbells have amassed a five acre collection of 5 and 15 gallon material. They will broker plants for the landscape trade and are working on a display garden in Mendocino County.

How to get there: Call for directions.

General information: Open to wholesale customers only, Monday through Friday, 7:30 am - 4:30 pm. Retail customers may visit with landscape professional.

County: Sonoma

MARCA DICKIE NURSERY

P. O. Box 1270
Boyes Hot Springs 95416
(707) 996-0364
Marca Dickie

Retail and Wholesale

Plant Specialties: 115 cultivars of grafted Maples, mostly of Acer palmatum, mostly for landscape use, some dwarfs. 1 gallon to 36" boxes.

History/description: During her ornamental horticultural studies, Marca Dickie discovered that she had a knack for propagation and grafting - perhaps due to her previous art training. Choosing to go into business for herself, she succumbed to the allure of maples, nice to work with, so forgiving, and so many varieties to stimulate her work. She maintains a display garden, does contract growing and will help design your garden. She suggests you plan also to see the interesting, labelled trees in Sonoma Town Square and the native plants in the Mission garden.

How to get there: Call for directions.

General information: Open by appointment only They will ship one gallon plants only in mid-winter. Plant list is available. On special occasions they will give talks.

County: Sonoma

McALLISTER WATER GARDENS

7420 St. Helena Highway (Highway 29)
Yountville 94599
(707) 944-0921, FAX (707) 944-1850
Vicky and Walt McAllister

Retail

Plant Specialties: Aquatic plants - Water Lilies (50 varieties), Water Iris (20 varieties), Bog plants (100 varieties), flowering Perennials (125 varieties).

History/description: The McAllisters started this business because they wanted to provide a local source for container-grown aquatic plants. Catalogue-sales companies ship their plants bare root which diminishes their survival rate. The McAllisters are the largest seller of aquatic plants in Northern California. They supply 75-80 nurseries with aquatic plants and produce a commercial newsletter. Contained water gardens offer a welcome and water-conserving relief for dry areas.

How to get there: The nursery is located on Highway 29, two miles north of Yountville in the mid Napa Valley.

General information: Open March through September, Friday, Saturday, Sunday, 9 am - 4 pm. They ship mail-orders, but primarily they encourage people to visit the nursery or buy from local suppliers. Price list is available.

County: Napa

MINIATURE PLANT KINGDOM

4125 Harrison Grade Road
Sebastopol 95472
(707) 874-2233
Becky and Don Herzog

Retail and Wholesale

Plant Specialties: Miniature Roses, Bonsai starters, Dwarf Conifers, Alpines, Dwarf Perennials, Rock Garden plants.

History/description: The Herzog's decision to start a nursery of miniature plants was based, in part, on the sweet memory of a high school project to create a miniature rose greenhouse. Their 2.5 acre nursery perched on a mountain top and surrounded by apple orchards is the result of more than a quarter century of hard work. They are well-connected with Sonoma county growers and generous with their knowledge.

How to get there: Take Highway 101 north to Highway 116 (about 5 miles north of Petaluma). Go left (west) on Highway 116 through Sebastopol. Left on Graton Road for 5.5 miles. Right on Tanuda Road at crest of mountain. At end of block go right for one block, then left. Their driveway is on the right in .6 miles. Sign at driveway says Occidental Nursery. At end of driveway, they are on the left.

General information: Open everyday but Wednesday, 9 am - 4 pm; Sunday, noon - 4 pm. Not open for wholesale on Saturday or Sunday. They ship plants and have a mail-order catalogue available for $2.50.

County: Sonoma

MOMIJI NURSERY

2765 Stony Point Road
Santa Rosa 95407
(707) 528-2917
Sachi and Mike Umehara

Retail

Plant Specialties: Grafted varieties of Japanese maples, exclusively Acer palmatum.

History/description: Although only open to the public since 1986, Momiji Nursery represents the Umehara's lifelong interest in Japanese maples. Their specimen plants are exquisite, and they have over 70 cultivars of Acer palmatum.

How to get there: Take Highway 101 north to Yolanda/Hearn exit. Turn left onto Santa Rosa Avenue. At signal light, turn left onto Hearn Avenue and go over freeway. Stay on Hearn Avenue for 1.6 miles. Go left at stop sign onto Stony Point Road.

General information: Open by appointment only. They deliver throughout the San Francisco Bay Area.

County: Sonoma

MOSTLY NATIVES NURSERY

27235 Highway One,
P. O. Box 258
Tomales 94971
(707) 878-2009
Margaret Graham, Walter Earle

Retail and Wholesale

Plant Specialties: Drought-tolerant plants, especially coastal California native plants. Native Trees (1 gallon), Perennials (4" pots), Native Bunch Grasses (1 gallon and 4" pots).

History/description: This partnership has been operating from the heart of downtown Tomales since 1984. With horticulturally-minded owners and a location near the sea, Mostly Natives is in a good position to supply coastal gardens in west Marin and Sonoma counties with plants that will really grow. They even know which deer will eat what.

How to get there: Highway 101 north to Petaluma/Washington Street exit. Continue west to Two Rock. Follow signs to Tomales.

General information: Open Tuesday through Saturday, 9 am - 4 pm; Sunday 11 am - 4 pm. They deliver plants to the North Bay, produce a plant list, and give talks. Mail order catalogue, "West Coast Natives", and current update is available for $3.

County: Marin

MUCHAS GRASSES

P. O. Box 683
Occidental 95465
(707) 874-1871, (707) 586-FARM
Bob Hornback, Michael Golas

Wholesale

Plant Specialties: Ornamental Grasses, Native grasses, Grasslike plants, 100 varieties and growing, 2" pots to 5 gallon cans.

History/description: Although this nursery was started in 1991, both owners have managed local nurseries and have backgrounds in landscape design and ornamental horticulture. More importantly, they have been collecting choice specimens for years. Their display garden illustrates landscape uses of grasslike plants and they give first-rate talks on their favorite subject. Bob Hornback continues to teach design and horticulture at Santa Rosa Jr. College and does garden installations.

How to get there: Call for directions.

General information: Open by appointment only. Plant list available for S.A.S.E. They deliver locally.

County: Sonoma

NEON PALM NURSERY

3525 Stony Point Road
Santa Rosa 95407
(707) 585-8100
Cindy and Dale Motiska

Retail and Wholesale

Plant Specialties: Palms, Cycads, Conifers and Subtropical plants.

History/description: The Motiskas have plants available in all sizes - from seedlings to mature specimens over 40 feet tall. They can transplant large specimens by crane. Since 1984 they have made an important contribution to the diversification of palm species. Allow time to visit their two acre botanical garden. Guided tours are on Sundays.

How to get there: Highway 101 to Santa Rosa. Take Todd Road exit west 1.5 miles to Stony Point Road. Right at stop light onto Stony Point. Nursery is .3 miles on the left.

General information: Open Tuesday through Saturday 10 am - 5 pm; Sunday 12 - 5 pm. Guided tours on Sunday. They ship and deliver to northern California. Catalogue available for $1. Slide talks available for interested groups.

County: Sonoma

NORTH COAST RHODODENDRON NURSERY

P. O. Box 308
Bodega 94922
(707) 829-0600
Parker Smith

Retail and Wholesale

Plant Specialties: Rhododendrons and Azaleas, species and hybrids, specializing in rhododendrons for mild climates (semi-tropical) and deciduous azaleas.

History/description: Lured west in the volatile 1960s to attend UC Berkeley, Parker Smith earned his degree in Landscape Architecture. In conjunction with his professional practice, he opened the nursery in 1983. Rhododendron 'Winter Peach', named by him, has become the main source of rhododendron hybrids in California (35 varieties in a wide range of colors and sizes). He can help design your garden and will talk to groups about rhododendron hybridizing.

How to get there: Call for directions.

General information: Primarily mail-order, the nursery is open on occasional weekends in March, April and May, by appointment only. He will ship plants and has an established mail order business. Catalogue available for $1.

County: Sonoma

OASIS

84 London Way
Sonoma, 95476
(707) 996-8732
Shirley and Kermit Puls

Retail and Wholesale

Plant Specialties: Cacti and Succulents.

History/description: A full-time nursery since 1983, the breadth of selection at Oasis reflects the Puls' 25 year collecting hobby that admittedly "got out-of-hand". They started propagating from seed when so many plants desired for their collection proved to be unavailable. When they realized their success with seeds had spawned 600 different varieties, they began to think sales. All plants are propagated on the premises and grown locally under polyethylene or shade cloth, producing a full winter dormancy; this, we are told, produces a hardier plant with better color and spination.

How to get there: Call for directions.

General information: Mostly Wholesale. Retail business limited to the serious collector, by appointment only, March through October.

County: Sonoma

PETITE PLAISANCE - ORCHIDS

P. O. Box 386
Valley Ford 94972
(707) 876-3496, FAX (707) 876-3496
Jim Hamilton, Ron Ehlers

Retail and Wholesale

Plant Specialties: Orchids - rare and unusual. Orchid supplies.

History/description: In 1981 these orchid hobbyists became partners in a nursery nestled in the rolling hills of Sonoma county, two miles from Bodega. They tramp the globe to collect and propagate orchids to preseve endangered species. Travels take them to Peru, Ecuador, Mexico and Southeast Asia. Orchids comprise the largest number of flora on the planet - some 35,000 species have been identified - and these guys want them all! They also have some pretty interesting tales to tell about orchid pollination.

How to get there: Call for directions.

General information: Open by appointment only. They ship worldwide and have a plant list available. They will give talks and demonstrations to groups.

County: Sonoma

PLANTS FROM THE PAST

P. O. Box 372
Bolinas 94924
(415) 868-1885
Sarah Hammond

Retail

Plant Specialties: Perennials, Mediterranean plants.

History/description: Impeccably credentialed in the world of horticulture, Sarah Hammond apprenticed at several well known English gardens. She worked for three years with Marshall Olbrich at Western Hills Nursery and was the founding director of the nursery at Smith & Hawken for seven years. British-born, she shares her country's passion for new plant discovery. Each year she returns home to search for plants which combine the "English" look with a tolerance for our dry climate. Lavatera 'Barnsley' was one of her good finds. Her display garden will inspire your purchase.

How to get there: Call for directions.

General information: Open by appointment only. Plant list is available for S.A.S.E. and her talks have enlightened many groups.

County: Marin

SHRUB GROWERS NURSERY

3260 Redwood Road
Napa 94558
(707) 226-8985
Barbara and Arthur MacDonald

Retail (limited) and Wholesale

Plant Specialties: California native plants, especially Bays, Madrones, Oregon Ash and shade-loving Ground Covers. Drought-tolerant plants.

History/description: Arthur MacDonald started this business in 1964 under the name of Woodside Gardens. Trained as an architect, he saw great possibilities in opening a nursery which could offer full design services. In 1986 the nursery's name was changed to reflect its concentration on larger water-conserving specimens. Nearby horticultural attractions include Old Bale Mill Park and Beaulieu Vineyards in Rutherford.

How to get there: Take Highway 101 to Highway 121 towards Sonoma and Napa. Follow signs to Napa. Take Highway 29 towards Napa. Left on Redwood Road. Nursery is in 3 miles.

General information: Open by appointment only. Deliveries based upon size of order. Plant list available.

County: Napa

SKYLARK WHOLESALE NURSERY

6735 Sonoma Highway
Santa Rosa 95409
(707) 539-1565
John Farmar-Bowers

Wholesale

Plant Specialties: California native plants, Mediterranean plants, with an emphasis on plants which are cold-hardy and drought-tolerant.

History/description: Skylark Nursery has been synonymous with California native and mediterranean plants since 1951. A pioneer in the production of appropriate California flora, this nursery's mission has remained essentially the same since new owners bought it in 1975. They offer tours of their 113 acres in the Valley of the Moon to pre-scheduled groups. Lisbeth Farmar-Bowers is in charge of the Sonoma Flower Company located next door. The Saturday sales include her dried arrangements and gift materials usually sold through retail stores and Gardener's Eden.

How to get there: Nursery located on the north side of Highway 12 at Oakmont Drive intersection. Six miles east of Highway 101. Sixteen miles west of Sonoma.

General information: Open for three Saturdays in the fall. Wholesale only all other times, but they welcome you with your designer or contractor. They will ship throughout California.

County: Sonoma

SONOMA ANTIQUE APPLE NURSERY

4395 Westside Road
Healdsburg 95448
(707) 433-6420
Caroline and Terry Harrison

Retail and Wholesale

Plant Specialties: Antique Apples, Pears, Peaches and Plums. Of particular interest are espaliered fruit trees, trained on trellises for two years and ready to bear fruit.

History/description: Located in the hills west of Healdsburg, this nursery is certified organic. The Harrisons now grow 100 varieties of antique apples, pears, peaches and plums. Look out for their mid-October apple tasting. You never have to settle for Golden Delicious again. The Harrisons will give talks to interested groups.

How to get there: Highway 101 north to Central Healdsburg exit. Go left at first stop at Mill Street. Mill Street becomes Westside Road. Nursery in 4 miles. From 101 south, take Westside exit. Go right.

General information: Open January 15 - March 30, Tuesday through Saturday, 9 am - 4:30 pm. Rest of the year by appointment only. They ship bare root, deliver container trees to the Bay Area, and have a catalogue available for $2 (refundable with purchase).

County: Sonoma

SONOMA HORTICULTURAL NURSERY

3970 Azalea Avenue
Sebastopol 95472
(707) 823-6832
Polo de Lorenzo, Warren Smith

Retail and Wholesale

Plant Specialties: Rhododendrons, Azaleas, unusual Trees, Shrubs and companion plants.

History/description: Anyone looking for a real springtime show should plan a visit to Sonoma Hort's five acres of display gardens. In addition to the fields of rhododendron, azalea and other ericaceous blossoms, there is an impressive laburnum walk and maple alley. Visitors will also enjoy the owner's eighteen years of accumulated knowledge.

How to get there: Take Highway 101 north to Highway 116 west. Go 4 miles west on Highway 116. Left on second Hessel Road. After one block, right on McFarlane. After six blocks, right on Azalea Avenue.

General information: Open everyday in March, April and May. Open Thursday - Monday, 9 am - 5 pm, from June through February. Catalogue is available for $2.

County: Sonoma

A STICKY BUSINESS

110 Liberty Road
or P. O. Box 743
Petaluma 94953
(707) 795-3185
Allan Leroy

Retail and Wholesale

Plant Specialties: 1,000 varieties of Cacti and Succulents, in 2" pots to 20" containers, also dish gardens. 400 varieties of Sempervivum and Sedum.

History/description: Ever since his days as a horticulture student at San Francisco City College studying under Jack Napton, Allen Leroy was intrigued by cacti and succulents. In 1978 after a brief stint in landscape maintenance and interior plantscapes, he leased a former chicken coop, converted it to a greenhouse, and went into the nursery business. Now he has more than 200,000 plants in eleven poly houses and outdoor growing space. Consider a visit also to nearby General Vallejo's Old Petaluma Adobe.

How to get there: From Highway 101 north, take Penngrove exit, the last exit in Petaluma. Go left over freeway. Go right onto Stony Point North for one mile. Left onto Pepper Road for one mile. Left onto Liberty Road.

General information: Open Saturday and Sunday, 10 am - 5 pm. Wholesale, weekdays, by appointment only. Deliveries of larger specimens can be arranged.

County: Sonoma

UP SPROUT GERANIUMS

237 Montgomery Street
Sebastopol 95472
(707) 829-6654
Sylvia Musso

Retail (Sebastopol); Wholesale (Livermore, 510-443-7638)

Plant Specialties: 1,000 varieties of Pelargoniums (mini, dwarf, fancy-leaved, scented) and other members of Geraniaceae family (no hardy geraniums), Ivy Geraniums (12 varieties).

History/description: In 1978 while recovering from a bad accident sustained while training her Tennessee Walking Horse, Sylvia Musso saw an ad for a 4'x7' greenhouse kit. To fill the greenhouse, she decided to grow something which was not easily available, something which would require her expected year of recuperation to mature and something which would use little heat. Hence pelargoniums. A full-time business since 1981, she now has 20,000 plants under polyethylene, most in her wholesale location in Livermore. Incredibly, she still finds time to show her championship mule and ballroom dance competitively four times a week.

How to get there: Highway 101 north to Highway 12 (Bodega Highway). Right on Montgomery.

General information: Retail operation in Sebastopol open Saturday, 9 am - 4 pm; Sunday, 9 am - 3 pm. Wholesale location in Livermore open by appointment only. They will ship mail orders. Plant list available for Pelargonium 'Americana' and 'Eclipse' series only (S.A.S.E.).

County: Sonoma

North Bay

URBAN TREE FARM

3010 Fulton Road
Fulton 95439
(707) 544-4446
Stanley Bradbury, Ruth Martin; Raphael Alvarez, Manager

Retail and Wholesale

Plant Specialties: Specialists in big Trees, Broad-leaved evergreens, Flowering trees, Fruit trees, Palms, Conifers, Shrubs. Sold as 5 gallons and up. Height up to 15'.

History/description: Former farmer, Stanley Bradbury, worked for years for the southern California Agricultural Department while growing Chritmas trees at home for sale. Originally more of a hobby than enterprise, the growing of conifers expanded to broad-leaved trees which led to a full-fledged wholesale business. Twenty-five years after his first Christmas tree sale and after his retirement from the Ag Department, he opened a retail nursery. At this new location since 1991, Urban Tree Farm now grows 30,000 trees on 20 acres and includes a display garden.

How to get there: From Highway 101, go west on River Road, then south on Fulton road for half mile.

General information: Open Monday through Saturday, 8 am - 5 pm; Sunday, 10 am - 5 pm.

County: Sonoma

URMINI HERB FARM

1292 Hurlbut Road
Sebastopol 95472
(707) 829-0185
Larry Urmini

Retail

Plant Specialties: Culinary and medicinal Herbs.

History/description: Larry Urmini comes from a long line of great Italian cooks who prize their fresh herbs. He decided to grow herbs in 1981 to have fresher ones available for his large Sonoma County pasta business. He now grows and sells 100 varieties of herbs for both cooking and healing.

How to get there: Highway 101 north to Highway 116 west towards Sebastopol. Go right on Hurlbut.

General information: Open by appointment only. He will gladly give talks to groups.

County: Sonoma

VINTAGE GARDENS

3003 Pleasant Hill Road
Sebastopol 95472
(707) 829-5342
Gregg Lowery

Retail

Plant Specialties: 350 varieties of Roses, specializing in the old, unusual and hard-to-find.

History/description: Always interested in horticulture, Gregg Lowery took a while to get there. This former high school teacher, social worker, and offset printer started in the nursery and landscape design business in 1982. He had a cut-flower farm in San Francisco, moved to Sebastopol and founded Vintage Gardens in 1985.

How to get there: Write to them requesting to be on mailing list for their Open Garden Weekend. They will send you the date and directions.

General information: Open only one weekend a year, in May during peak of the rose season. Rest of the year, their business is strictly mail-order. Mail-order catalogue is available for $3.

County: Sonoma

WESTERN HILLS NURSERY

16250 Coleman Valley Road
Occidental 95465
(707) 874-3731
Maggie Wych

Retail

Plant Specialties: Large assortment of unusual Perennials, Shrubs, Vines and Trees.

History/description: Back in 1960 when there were few responsible thinkers about appropriate horticulture in California, Marshall Olbrich and Lester Hawkins started to propagate plants for their own extensive garden. Western Hills has been on the cutting edge of plant introductions ever since. The knowledge shared by these beloved horticulturists spawned a whole generation of specialty growers and plant hobbyists. With the recent death of Marshall Olbrich, the nursery was willed to Maggie Wych who had worked with Marshall for eleven years. Western Hill's three acre display garden is chock-full of mature specimens of rare and unusual plants. They are located in the redwood country, 60 miles north of San Francisco.

How to get there: Highway 101 North to Rohnert Park, Highway 116 west exit. Follow Highway 116 through Sebastopol to Occidental Road. Left on Occidental for 5-6 miles to village of Occidental. Turn right on Coleman Valley Road for about a mile.

General information: Open Thursday through Sunday, 10 am - 5 pm. In December and January, by appointment only.

County: Sonoma

WILDWOOD FARM

10300 Sonoma Highway
Kenwood 95452
(707) 833-1161
Ricardo and Sara G. Monte

Retail and Wholesale

Plant Specialties: California native plants, cold-hardy Perennials, Conifers, large-sized grafted Japanese Maples. Many unusual, pendulous, contorted Trees.

History/description: The evolution of a plantsman as lived by Ricardo Monte: start with a degree in history from UC Berkeley, become a fledgling dress designer, shift to garden design, fall in love with plants, start locating and growing rare and unusual specimens for use in your own landscape projects, sell some, collect even more, then open a nursery on fifteen rural acres. He continues to do residential garden design.

How to get there: Take Highway 101 north to Highway 12 east. Nursery is in Kenwood on the west side of the highway.

General information: Open March through October, everyday but Monday and Wednesday, 9 am - 4 pm. Winter dates, everyday but Monday, Wednesday and Sunday, 9 am - 4 pm. They make local deliveries only to Sonoma and Marin.

County: Sonoma

YA-KA-AMA NURSERY

6215 Eastside Road
Forestville 95436
(707) 887-1541
Ya-Ka-Ama Indian Education & Development Program;
Ria Young, Nursery Manager

Retail and Wholesale

Plant Specialties: California native plants, Herbs, Ornamental Grasses.

History/description: This non-profit enterprise raises money for its education center and vocational training program for Native American youth through the sale of California native plants and grasses. It has the largest selection of ornamental grasses in northern California. It is also developing as a horticultural center with sales, classes, herb garden, and demonstration gardens which illustrate the 11 plant communities in Sonoma County. A very effective and worthwhile effort.

How to get there: Highway 101 north to River Road exit north of Santa Rosa. Go west on River Road for 6 miles. Go right on Healdsburg-Trenton Road for 1 mile to end. Go left on Eastside Road. Ya-Ka-Ama is 60 yards on the right.

General information: Open Monday - Friday, 9 am - 5 pm; Saturday by appointment. They deliver plants to Sonoma and Marin counties and soon to the East Bay. They will give talks to interested groups.

County: Sonoma

HORTICULTURAL ATTRACTIONS

Green Gulch Farm

Off Highway 1, between Mill Valley and Muir Beach, or Star Route
Sausalito 94965
(415) 383-3134 (office)

History/description: Green Gulch is a Zen Buddhist center and organic farming community. Produce from their 15 acre farm appears in the best restaurants and health food stores. The smaller garden, separating the living quarters from the farm and sea beyond, includes rose arbors, a formal herbal circle, espaliered fruit trees and masses of flowers. Saturday classes are offered on such subjects as composting and wreath making. A robust Garden Volunteer program gives everyone a chance to be part of this serene and successful tending of the earth.

General information: Open everyday.

Plant sales: Their annual plant sale is in April, although some plants are always available for self-service sales.

County: Marin

Luther Burbank Home & Gardens

Corner of Santa Rosa & Sonoma Avenues,
P. O. Box 1678
Santa Rosa 95402
(707) 524-5445

History/description: It was here that Luther Burbank's experiments produced the Plumcot, the fast-growing 'Paradox' Walnut, the spineless Cactus and, of course, the 'Shasta' Daisy. His home and garden were given by his family to the city of Santa Rosa and are now a registered landmark. The 1.5 acre gardens were revitalized in 1991 to include signage and demonstration beds, worth a visit even when the home is closed. Tours, required for entry to the house, are given Wednesday through Sunday, from the first Wednesday in April to mid-October. There is no charge to visit the garden; house tours cost $2. This guide to plant specialties would be incomplete without a mention of California's greatest plant specialist.

General information: Garden is open everyday during daylight hours. Home open 10 am - 3:30 pm.

Plant sales: Plants are occasionally sold on site.

County: Sonoma

Other Attractions:

Korbel Champagne Cellars Garden
13250 River Road
Guerneville
(707) 887-2294

Marin Art and Garden Center
30 Sir Francis Drake Blvd.
Ross 94957
(415) 454-5597

Ten acres of old trees and memorial gardens; venue for shows and plant sales of a few garden groups.

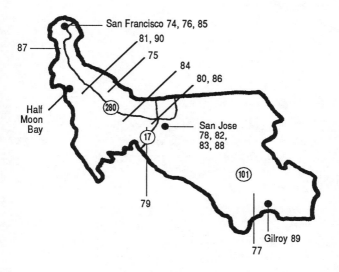

SAN FRANCISCO PENINSULA

SAN FRANCISCO
SAN MATEO
SANTA CLARA

ARBOR & ESPALIER CO.

201 Buena Vista Avenue East
San Francisco 94117
(415) 626-8880
John Hooper

Retail and Wholesale

Plant Specialties: Fruit trees, especially unusual and antique apples and pears. Wide variety of espalier forms, ranging from the formal, palmette verrier style to informal, fan-shaped designs. Plants sold bareroot and in 15 gallon containers.

History/description: Jock Hooper started this business in 1985 to offer a wide range of fruit trees in forms suitable for small gardens. Arbor & Espalier is presently the largest nursery devoted exclusively to espaliers in the country and has a branch nursery in Essex, MA. Many of their uncommon fruit trees result from a joint venture with Sonoma Antique Apple Nursery.

How to get there: The San Francisco Display Garden at 201 Buena Vista East is near the intersection of Haight and Baker Streets. To reach the Healdsburg nursery (4395 Westside Road), refer to Sonoma Antique Apple Nursery listing.

General information: Open by appointment only. Arbor & Espalier ships nationwide. Slide talk available for groups. Catalogue $2, applied toward purchase.

County: San Francisco

BAYLANDS NURSERY

1165 Weeks Street (nursery location),
2835 Temple Court (mailing address)
East Palo Alto 94303
(415) 323-1645
Day and Yuki Boddorff

Retail and Wholesale

Plant Specialties: California native plants, including Ceanothus, Arctostaphylos, western Redbud. Ornamental Grasses (over 50 varieties). Unthirsty Ground Covers, Perennials, Trees, and Shrubs, including Leucodendron and Protea.

History/description: Baylands is an outgrowth of Day Boddorff's desire to have more interesting plants available for his landscaping business. His knowledge of plants has been enriched by his far-flung past. Five years of living in South Africa studying botany and soils gave him more than a passing familiarity with proteas. A Masters Degree from the U. of Florida in Agronomy, studying soils and pastures, explains his fascination with grasses. His move to California explains the rest. Baylands will special order plants at no extra charge. Site consultation, design and installation available.

How to get there: Baylands Nursery is located one block west of the Bay Road/Pulgas Avenue intersection in East Palo Alto. Call for directions.

General information: Open Tuesday through Saturday, 8 am - 4 pm; Mondays by appointment. Delivery to most of San Mateo and Santa Clara counties with minimum order (Please inquire). Mail order service available.

County: San Mateo

BONSAI GARDEN

246 - 28th Avenue
San Francisco 94121
(415) 752-6436
Elinor Carlton

Retail

Plant Specialties: Large, potted Bonsai.

History/description: Elinor Carlton was hooked on bonsai from the moment she saw pictures of them as a child in Mississippi. Eager to exchange her hot summers for our cool fog, she moved west in 1954 and bought her first bonsai. One plant is never enough and in 1977 her passionate hobby became a business. She purchases older (30 years and up), grafted specimens from expert growers and painstakingly trains each one herself.

How to get there: Located in San Francisco's Seacliff district, between California and Lake Street.

General information: Open Wednesday through Friday, 12:30 - 4:00 pm, or by appointment. A great lecturer, she can unravel the mysteries of bonsai for any group.

County: San Francisco

BRIARWOOD NURSERY

1910 E. San Martin Avenue
San Martin 95046
(408) 683-2632
Craig Pierce

Wholesale

Plant Specialties: Ornamental Shrubs and Herbs. Herbs are available by mailorder only.

History/description: Craig Pierce decided growing plants was much more interesting than working in a retail nursery. His training as a botanist has found successful application. He currently supplies the herbs for Gardener's Eden catalogue sales. Hecker Pass Family Theme Park with its fabulous trees is nearby off Hecker Pass Road (Highway 152).

How to get there: Call for directions.

General information: Open to wholesale customers only, by appointment. Public may come with landscape professional. He will deliver plants locally.

County: Santa Clara

C. H. BACCUS

San Jose 95117
(408) 244-2923
Charles Baccus

Retail and Wholesale

Plant Specialties: Bulbs, Calochortus (50 species) and other California native bulbs.

History/description: Charles Baccus, fresh out of horticulture school in 1977, started a part-time business in native orchids. This eventually became a native bulb business. Now retired from his regular job, he hopes to establish growing grounds in other states. He has a display garden, (best to see in bulb blooming season) and he will do some contract growing.

How to get there: Call for directions.

General information: Open by appointment only. Plant list available for S.A.S.E.

County: Santa Clara

CARMAN'S NURSERY

16201 E. Mozart Avenue
Los Gatos 95032
(408) 356-0119
Ed Carman

Retail and Wholesale

Plant Specialties: Hostas, Rock garden plants, Bonsai starters, other unusual Perennials and 7 named Wisteria cultivars. Hypertufa troughs for alpine plants.

History/description: The history of Carman's reflects the evolution of the nursery trade in northern California, albeit Carman's has always been a bit ahead of the times. The business, started by Ed's father in 1937, began with bedding plants, then shifted to selling herbs in pots to retailers. In the 1960s they began to focus on perennials. By the 1970s they were determined to specialize in unusual perennials from around the world. In his present location since 1970, Ed Carman is known as one of the best sources for interesting perennial material. Nearby horticultural attractions include Hakone Gardens in Saratoga, Japanese Tea Garden in San Jose, Vasona Park in Los Gatos and Villa Montalvo in Saratoga.

How to get there: Take Highway 101 south to Highway 17 west. Exit Camden. Go left on Camden. Right on Bascom. Right (.9 miles) on Mozart.

General information: Open Tuesday through Saturday, 9 am - 5 pm.

County: Santa Clara

CHRISTENSEN NURSERY COMPANY

16000 Sanborn Road
Saratoga 95070
(408) 867-4181
Jack Christensen; Albert Sotelo, Manager

Retail and Wholesale

Plant Specialties: California native plants, including Ceanothus, Toyon, Myrica, Fremontodendron, Carpenteria, Madrone, Arbutus 'Marina', Manzanita, Oaks, Maples, Dogwoods. Camellia species, including japonica, sasanqua. General line of Ornamental Trees and Shrubs.

History/description: This family operation was founded by Jack Christensen's father in 1938, moving to Belmont in 1947 and to Saratoga in 1960. Though the business is almost entirely wholesale, they welcome knowledgeable plant enthusiasts.

How to get there: From Saratoga, go west on Route 9 for two miles. Opposite Saratoga Springs picnic area, go left on Sanborn Road.

General information: Open Monday through Saturday, 8 am - 4:30 pm.

County: Santa Clara

FARWELL'S NURSERY

13040 Skyline Boulvard
Woodside 94062
(415) 851-8812
Mrs. Everett E. Farwell, Jr.

Retail

Plant Specialties: Rhododendrons and Azaleas, mostly ungrafted hybrids, some species and unusual varieties, also large specimen sizes, all field-grown

History/description: In 1946 the Farwells decided to go into business for themselves, choosing to pursue his avocation of rhododendron hybridizing. At that time the only rhododendrons available in California were large-sized plants from Oregon. The Farwells were the first in California to sell two year old, field-grown specimens. Farwell's has a well-established display garden on-site. Their location at 2,000' elevation encourages the culture of cold hardy plants. Mrs. Farwell will give talks to interested groups.

How to get there: From Highway 101, go west on Highway 92. Go south on Skyline Boulevard for 4.5 miles.

General information: Open March through June, Friday through Tuesday, 10 am - 4 pm. August through February, on weekends, 10 am - 4 pm. Closed in July.

County: San Mateo

INSTANT OASIS

Nursery location in Campbell,
3079 Greentree Way (office)
San Jose 95128
(408) 241-6084
FAX (408) 374-3505, enter 99 when answering machine comes on.
Ann Kuta

Retail

Plant Specialties: Aquatic plants - over 100 varieties of Bog and Upright plants, 50 types of hardy and tropical Lilies. Complete line of water garden components.

History/description: While living in Colorado thirty years ago, Ann and Don Kuta embarked upon a lifelong interest in designing aquatic gardens. Sensing a need to help customers create complete water gardens, Don began designing custom filters and sharing his many years of expertise building ponds. They can supply you with anything necessary for a tub garden to a total back yard transformation.

How to get there: Call for directions. The nursery in Campbell is one block from Winchester Boulevard, but hard to find due to the commercialization of the area.

General information: Open by appointment only. Seasonal hours.

County: Santa Clara

MARYOTT'S IRIS GARDEN

1069 Bird Avenue (Display Garden),
1073 Bird Avenue (Office)
San Jose 95125
(408) 971-0444
William R. Maryott

Retail

Plant Specialties: Bearded Iris - bloom-sized rhizomes

History/description: What began as a hobby - the growing and hybridizing of tall bearded iris - became a business in 1977. Maryott's display garden in the Willow Glen area of San Jose is nearly a full acre with over 550 varieties.

How to get there: Take Bird Avenue exit off of Highway 280. Travel 8 blocks south. 1069 Bird is before you get to Willow Street.

General information: Open during bloom season only (approximately mid-April through first week in May). Then mailorder only until August 15. Call for bloom status. They ship throughout the USA. Plant list is available for 2 stamps.

County: Santa Clara

NATIVE REVIVAL NURSERY

855 Emory Avenue
Campbell 95008
(408) 374-7349
Erin and Dan O'Dougherty

Retail and Wholesale

Plant Specialties: California native plants from San Francisco Bay region.

History/description: For several years the O'Doughertys supplied home-grown plants to a landscape contractor friend. These plants were so well received, they went into business in 1992. Trained in ornamental horticulture at Foothill College and by Yerba Buena Nursery, they currently contract grow for water districts, parks, individuals, and, of course, their friend. They offer site-specific consultations and landscape installation and mantenance services. They will also give lectures and talks.

How to get there: Take 101 to Highway 17 west to Camden/San Tomas exit. Go right on Winchester South. After two blocks go right on Sunnyoaks. Right on Emory.

General information: Open Friday, 9 am - 5 pm and Saturday, 9 am - 4 pm and by appointment. Wholesale customers by appointment. Plant list is available and they will give talks and tours to interested groups.

County: Santa Clara

ROD MCLELLAN CO. ACRES OF ORCHIDS

1450 El Camino Real
South San Francisco 94080
(415) 871-5655

Retail and Wholesale

Plant Specialties: Orchids, many hybrids introduced, including the introduction of hybrid genus, McLellanara.

History/description: Rod McLellan's grandfather started this business as a dairy in 1888. A family owned business, Rod McLellan Co. has been growing orchids for over 50 years. Their 35 acre site in South San Francisco is well worth a visit for its wealth of greenhouses and propagation laboratories. Daily tours at 10:30 and 1:30 explain the intricacies of germination and meristem culture. They are open for retail sales on Fridays at the 120 acre growing ground in Watsonville. They have a boarding service for out-of-bloom orchids and offer services and classes in repotting and orchid care.

How to get there: From Highway 280 take Hickey Boulevard Exit. Follow Hickey to El Camino. Turn right at El Camino. Nursery 100 yards on right.

General information: Retail is open everyday, 8 am - 5 pm. Wholesale, Monday - Friday, 8 am - 5 pm. They ship nationwide. Free plant list is available.

County: San Mateo

SASO HERB GARDENS

14625 Fruitvale Avenue
Saratoga 95070
(408) 867-2135
Virginia and Louis Saso

Retail

Plant Specialties: Culinary, ornamental and medicinal Herbs. An especially noteworthy collection of Salvias. They also propagate the multi-purposed Neem Tree, Azadirachta indica.

History/description: Louis Saso's interest in herbs stems from his earlier occupation as a wholesale produce merchandizer. This retirement hobby/business began in earnest in 1974. Now in addition to growing herbs, they sell wreaths and other handicrafts made from herbs. Saso keeps the entire South Bay herbally aware with regularly scheduled classes and workshops in herbal medicine, aroma therapy, wreath and garland making, herbal crafts, and herb garden design. Plan also to visit Hakone Gardens and Villa Montalvo, both in Saratoga.

How to get there: Take Highway 280 south to Highway 85 exit. Go right for 5 miles to Saratoga. Continue through signal light for one mile. Left on Fruitvale. From Highway 101 take Lawrence Expressway which ends at Saratoga Avenue. Go right on Saratoga Avenue for 2 miles. Left on Fruitvale.

General information: Open Thursday through Saturday, 9 am - 2:30 pm. They give talks to groups and offer free tours once a month on a Sunday, during April through August.

County: Santa Clara

SHELLDANCE NURSERY

2000 Highway One
Pacifica 94044
(415) 355-4845
Michael Rothenburg, Nancy Davis

Retail and Wholesale

Plant Specialties: Over 1,000 species and hybrid Bromeliads, including two of their introductions Aechmea 'Shelldancer' and A. 'Pacifica'. Orchids - Phalaenopsis, Oncidium, Dendrobium, Paphiopedilum

History/description: It really was a case of starting out "selling seashells down by the seashore". Hard to believe this incredible business began on San Francisco's Union Street selling Tillandsias in shells. They moved to Pacifica to expand their bromeliad collection. About the same time they met an orchid grower from Florida and went with him on collecting expeditions in tropical rainforests. Still focused on bromeliads, their attention turned to the propagation of rare and endangered species. They now include orchids for sale and in their magnificent rainforest display. They have the largest collection of bromeliads on the west coast, not all for sale at any one time, but all on view. The nursery is located at the trailhead for Sweeney Ridge, part of the Golden Gate National Recreation Area.

How to get there: Travelling south on Highway One, #2000 is on the left (east) side of road. There is a left-turn lane.

General information: Open Monday through Friday, 9 am - 4 pm. They will ship plants. Catalogue available for $1.

County: San Mateo

TREES OF CALFORNIA

Corner Flickinger and Falling Tree,
P. O. Box 13189
San Jose 95013
(408) 264-3663
Joe Arnaz

Retail and Wholesale

Plant Specialties: Huge (6'-14' boxed) Trees. Over 20 varieties plus most native Oak species, all hand dug and collected. Some unusual trees are 50-80 years old.

History/description: A bona fide tree hugger, Joe Arnaz actually does something about his passion. Constantly scouting for endangered trees, he arrives to dig just ahead of the bulldozer. He has worked as far away as Saudi Arabia and Chile, and has helped bring shade to important locations in California. Recent projects include the planting of 26,000 trees on a private estate near Woodside quarry, intended eventually to become an arboretum.

How to get there: Call for directions

General information: Open by appointment only.

County: Santa Clara

WRIGHT IRIS NURSERY

6583 Pacheco Pass (Highway 152)
Gilroy 95020
(408) 848-5991, FAX (408) 847-6495
Nathan Wright, Ean St. Clare

Retail and Wholesale

Plant Specialties: Iris, Canna.

History/description: These iris hobbyists started this business in 1981 for their retirement pleasure. They now have 1,000 varieties of iris (Louisiana, Spurian, and Siberian) and 59 varieties of canna in all colors. Their nursery is at the base of an old lake bed on the east side of the valley - an ideal growing ground. Goldsmith Seeds (flower fields can be seen from the road) and Mt. Madonna State Park are other horticultural attractions on Hecker Pass Road.

How to get there: Highway 101 south to Gilroy (Highway 152) exit. Nursery is 9 miles east of Gilroy on Highway 152.

General information: Open all year, 9 am - 5 pm. They ship their rhizomes UPS. Free plant list available for Californians, $3 elsewhere.

County: Santa Clara

YERBA BUENA NURSERY

19500 Skyline Blvd.
Woodside 94062
(415) 851-1668
Gerda Isenberg

Retail and Wholesale

Plant Specialties: California native plants (more than 500 species), native and exotic Ferns.

History/description: A student at horticulture schools in Germany, Gerda Isenberg was an early proponent of California native plantings. In her present location since the 1940s, she and her husband raised cattle before she developed an interest in the ferns growing on their property. Opening the nursery in 1952, she makes hard-to-get natives available to grateful customers. Her nursery location in a canyon on the west side of the coastal mountains provides a perfect setting for her large collection of ferns and native plants. A two acre demonstration garden allows visitors to see mature plantings of the nursery stock. To make certain her "Go Native" message gets to future generations, the nursery offers a ten-week internship program.

How to get there: Highway 101 or Highway 280 to Woodside Road Exit (Highway 84). Go west to Skyline. Left on Skyline for 4.5 miles. Right on Rapley Ranch Road for 2.2 miles. Keep to right for gate of nursery.

General information: Open everyday except major holidays, 9 am - 5 pm. Catalogue available for $1.

County: San Mateo

HORTICULTURAL ATTRACTIONS

FILOLI CENTER

Canada Road
Woodside 94062
(415) 364-2880 (Tour reservations), (415) 366-4640 (Friends)

History/description: Filoli is one of few remaining grand country estates in California. Now a National Trust for Historic Preservation property, Filoli was created in 1915-17 by William Bourne who inherited the Empire Mine at age seventeen. He plowed the same western energy into his garden as he did into the mine. Filoli's 16 acre garden is distinctly Californian with grand views of the coastal range, although it borrows from European garden tradition. Organized as a series of outdoor rooms, Filoli includes a woodland garden, rose garden, impeccably maintained formal gardens, and colorful seasonal displays. Friends of Filoli, the membership support group, organizes lectures and a Flower Show at the end of April every other year.

General information: House and garden open Fridays and some Saturdays for self-guided tours. Other days by reservation. Garden shop opens at 10 am from mid-February to mid-November.

Plant sales: Plants propagated at the garden are offered for sale in the Garden Shop.

County: San Mateo

Gamble Garden Center

1431 Waverly Street
Palo Alto 94301
(415) 329-1356

History/description: This turn-of-the-century Edwardian villa and garden was built by Edwin Gamble, grandson of the founder of Procter and Gamble. Coming to California to enter his son in Stanford, he fell in love with the area and built the house in 1902 for his daughter, Elizabeth F. Gamble, who really developed the grounds. Garden highlights include the wisteria, rose garden, cherry allee and an iris garden, the real favorite of Miss Gamble. Classes are held periodically on such topics as flower-arranging, topiary and drought-tolerant plants. A membership support group organizes a garden tour and luncheon in the Spring. Entrance to the garden is free; docent-led tours ($3) may be arranged by reservation.

General information: Garden is open anytime during daylight hours. The office and library are open from 9 am to noon on weekday mornings.

County: Santa Clara

SF League of Urban Gardeners

2088 Oakdale Avenue
San Francisco 94124
(415) 285-7584

History/description: Organized in 1983 as a non-profit, umbrella group for community and school gardens, S.L.U.G. offers classes on composting, organic gardening and integrated pest management. The display garden, "Garden for the Environment", at 7th Avenue and Lawton, demonstrates water-conserving plantings and good gardening techniques. About to double in size, the garden will add biodynamic demonstration beds, annual flower plantings, mock back yards, turf display and shade garden.

General information: Garden for the Environment is open everyday during daylight hours.

County: San Francisco

Strybing Arboretum & Botanical Gardens

Ninth Avenue and Lincoln Way Golden Gate Park
San Francisco 94122
(415) 661-1316, (415) 661-15114 (Library)

History/description: Officially opened in 1940, Strybing today has 70 acres containing 7,500 plants from all over the world. Plants are arranged in a series of gardens to educate and delight visitors, including a Garden of Fragrance, Succulent Garden, Conifer Garden, Cape Province Garden, Redwood Trail, and a Garden of California native plants. Strybing Arboretum Society, the Arboretum's membership support group, has a very active education program with lectures, workshops, tours, gardening clinics, ethnobotany programs, and children's activities. Strybing has a store and horticultural library which houses an impressive 14,000 volumes. Docent tours of the Gardens occur everyday at 1:30 pm and on the weekends also at 10:30 am.

General information: Open everyday during daylight hours. Library and Shop open 10 am - 4 pm.

Plant sales: Their big annual plant sale is held the first weekend in May; specialized plant sales take place on the second and fourth Saturdays of each month.

County: San Francisco

Golden Gate Park

c/o Friends of Recreation and Parks
McLaren Lodge, Golden Gate Park
San Francisco 94117
(415) 750-5105

Remarkable use of Mediterranean-climate plants to re-create English landscape garden setting. Special gardens include Strybing Arboretum (see separate listing), Shakespeare Garden, Conservatory of Flowers, Japanese Tea Garden.

Other Attractions:

Hakone Japanese Gardens

21000 Big Basin Way
Saratoga 95070
(408) 741-4994

Public garden (formerly private Stine garden), renovated in 1966, symbolic plantings, camellias, shaped trees, wisteria.

Overfelt Botanical Gardens

McKee Road at Educational Park Drive
San Jose 95127
(408) 251-3323

Botanical garden and Chinese cultural garden.

Prusch Farm Park

647 South King
San Jose 95116
(408) 926-5555

Small farm-style park with animals, fruit orchard and community gardens.

San Mateo Garden Center

605 Parkside Way
San Mateo 94403
(415) 574-1506

Over 40 horticultural groups hold their meetings, shows, sales, benefits, classes and workshops here.

San Jose Municipal Rose Garden
Naglee Avenue at Dana
San Jose 95126
(408) 287-0698

5,000 colorful and fragrant roses with redwoods and flowering trees.

San Mateo Japanese Tea Garden
Central Park
San Mateo
(415) 377-3340, (415) 377-3345

Meandering paths through serene plantings with koi pond, pagodas and tea house.

Villa Montalvo
15400 Montalvo Road
Saratoga 95070
(408) 741-3421

Country estate of Senator James Phelan, designed by John McLaren, inspired by European garden tradition.

■96 **Where On Earth!**

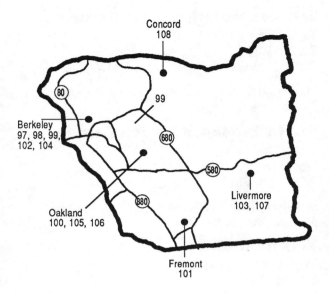

Concord
108

(80)

99

(680)

Berkeley
97, 98, 99,
102, 104

(580)

Livermore
103, 107

Oakland
100, 105, 106

(880)

Fremont
101

EAST
BAY

ALAMEDA
CONTRA COSTA

East Bay

ALPINES
1646 10th Street
Berkeley (510) 524-1969
John Andrews

Retail

Plant Specialties: Bulbs and Alpines from around the world, including Campanula, Penstemon, Erigonium, Primula.

History/description: An entomologist at UC Berkeley, John Andrews became interested in alpines while backpacking in the high country. He started gathering seed, joined the American Rock Garden Society, and was suddenly in the nursery business. All of his plants are propagated from seed, still collected by himself or trusted experts in the field, and grown in a special soil and gravel mix.

How to get there: From Highway 80, take University Avenue exit. Go east on University, then north on 10th Street.

General information: Open by appointment only Plant list is available for S.A.S.E.

County: Alameda

BERKELEY HORTICULTURAL NURSERY, INC.

1310 McGee Avenue
Berkeley 94703
(510) 526-4704
Paul Doty

Retail and Wholesale

Plant Specialties: Huge Vine selection, Rhododendrons, expecially Vireya, Maddenii. 300 varieties of Roses, including Old Roses. Subtropical Shrubs, unique, deciduous and asiatic Flowering Shrubs.

History/description: This business was started in 1922 and is now in its fourth generation of family ownership. The Dotys are responsible for the palm plantings at Hearst Castle in San Simeon and have introduced hundreds of fuchsia varieties to the Bay Area. The nursery has a "tropical" courtyard and rockery display, puts on workshops, and sponsors a yearly garden contest. Although few plants are grown on site, Berkeley Hort has a cadre of growers who contract grow specially for them. Their emphasis has always been on unusual and appropriate plant introductions.

How to get there: From Highway 80, take Albany exit. Go right on Hopkins. Right on McGee.

General information: Open April through October, everyday, 9 am - 5:30 pm. Winter hours, everyday but Thursday, 8:30 am - 5 pm. Plant list by subject, seasonally available for S.A.S.E.

County: Alameda

COPACABANA GARDENS

P.O. Box 323
Moraga 94556
(415) 254-2302
Lee Anderson, James Larsen

Retail and Wholesale

Plant Specialties: Unusual Subtropical and Drought-tolerant plants from the Southern Hemisphere. (South Africa, South America, Australia, Asia and the Philippines). Seedlings to 24" box.

History/description: Like many specialty growers, Copacabana Gardens started in response to the need for more interesting plant material for the many Bay Area micro-climates. Open only since 1989, their concentration on Southern Hemisphere flora has already yielded new plant possibilities for California gardens. This partnership also offers site consultations, design and landscape construction. They recommend you also visit Hacienda de las Flores in Moraga.

How to get there: Call for directions.

General information: Open by appointment only. Catalogue available for $2, although it lists only part of their stock. They ship small-sized plants and deliver orders over $150 throughout the Bay Area.

County: Contra Costa

THE DRY GARDEN

6556 Shattuck Ave.
Oakland 94609
(510) 547-3564
Keith Cahoon, Richard Ward

Retail

Plant Specialties: Mediterranean plants, Ornamental Grasses, and Cacti and Succulents for seasonally dry climate.

History/description: These self-confessed plant and garden maniacs are having a lot of fun growing and searching for weird and interesting plants. In the process, the permanent plantings surrounding their nursery have enlivened a previously dead corner of north Oakland.

How to get there: From Highway 80 east, exit on Ashby. Follow Ashby. Turn right on Shattuck. Nursery is 3 blocks on left. From Highway 580 west, take 24 east off ramp. Exit at Martin Luther King/51st Street. Turn right onto 51st, then left onto Shattuck. Nursery is 2 blocks after 3rd light on right. Only 10 minutes more to U.C. Botanical Garden and the Regional Parks Botanic Garden.

General information: Open everyday, 10 am - 6 pm. They deliver to the San Francisco Bay Area.

County: Alameda

FOUR WINDS GROWERS

P. O. Box 3538
Fremont 94539
(510) 656-2591
Don Dillon

Wholesale

Plant Specialties: Citrus

History/description: This major citrus growing business started in a Carmel backyard as a retirement hobby of Don Dillon's father. By 1949 he had contracted with a Ventura grower to produce plants for retail nurseries. The nursery was moved to Fremont in 1954. Four generations have since worked there. Don Dillon's activities have not been restricted just to plants; he is a past mayor of Fremont and past president of Association of Bay Area Governments. Don Dillon Jr. is currently president of the California Citrus Nurserymen's Society.

How to get there: Your landscaper must call for directions.

General information: Open to wholesale trade only. Visit with your landscaper. Catalogue and booklet on "How to Grow Citrus" available for S.A.S.E. They offer group tours by appointment and can arrange garden talks and slide shows by arrangement.

County: Alameda

GRASSLANDS NURSERY

2222 Third Street
Berkeley 94710
(510) 540-8011
Carol Williams, Manager

Retail and Wholesale

Plant Specialties: Primarily Native Grasses, Ornamental Grasses. Some Shrubs, Irises, California native plants.

History/description: Located on the Berkeley estuary, Grasslands was started in 1989 by Dr. Louis Truesdale, owner of neighboring American Soil Products. The same person who pioneered biochemically-improved soil mixtures and mulches should be able to work wonders with grass. They have a display garden, will custom grow and offer site-specific consultations.

How to get there: From Highway 80, take University Avenue exit. Follow University Avenue. Right on 6th Street. Right on Bancroft to end of street.

General information: Open Monday through Saturday, 7:30 am - 4 pm.

County: Alameda

JEFF ANHORN NURSERY

P. O. Box 2061
Livermore 94551
(510) 447-0858
Jeff and Renee Anhorn

Wholesale

Plant Specialties: Perennials, Woody Ornamentals, California native plants.

History/description: Another worthy graduate of the Landscape Architecture Department of UC Berkeley, Jeff Anhorn comes from prime horticultural stock. His uncle, Louis Gravello, owned many plant patents. One of these, for Escallonia 'Terri', financed the set-up costs for this nursery. He and his wife can help design your garden and offer site-specific consultations.

How to get there: Call for directions to their growing grounds in Sunol.

General information: Open to wholesale trade only, by appointment. Retail customers may visit with landscape professional. A plant list is available for S.A.S.E.

County: Alameda

MAGIC GARDENS

729 Heinz Avenue
Berkeley 94710
(415) 644-2351
Aerin Moore; Larry Greenwood, Nursery Manager

Retail

Plant Specialties: 750 varieties of Perennials, especially Salvias (21 varieties), Penstemons. Ferns, Roses, Aquatic plants, Rock Garden plants, Grasses.

History/description: Aerin Moore was taught good gardening practices by his grandfather and has been a gardener ever since. Impressed by what he could not find for his landscape clients, he started the nursery in 1982. Magic Gardens' new emphasis is on growing its own plant material. A recently acquired test garden in the Oakland Hills makes this possible. The beautifully landscaped rockery surrounding the nursery gives visitors the chance to see their purchases in a garden setting. They produce an informative newsletter and give classes and talks at the nursery.

How to get there: From Highway 80, take Ashby exit to first stop light. Go left on on 7th Street for three blocks. Go left on Heinz.

General information: Open Monday through Saturday, 9 am - 5 pm. Sunday, 10 am - 5 pm.

County: Alameda

MERRITT COLLEGE, ORNAMENTAL HORTICULTURE DEPARTMENT

12500 Campus Drive
Oakland 94619
(510) 436-2418

Retail and Wholesale

Plant Specialties: Perennials for shade and sun, focusing on rare finds, grafted Magnolias, Roses - old and new, Bulbs, Orchids, and Houseplants.

History/description: These student and faculty propagated materials earn money for the school's Ornamental Horticulture Department.

How to get there: Take Highway 580 east to 35th Avenue exit (which becomes Redwood Road). Go left over freeway up Redwood Road. Right on Campus Drive. Take first available left, go left again past Child Care Center. Go left at "T". Parking costs 50¢.

General information: Big sale in early May and October. Open thereafter until June, Mondays - Thursdays. Call first. Wholesale contact: Susan Ashley, (510) 524-3627.

County: Alameda

ORCHIDANICA

Oakland (510) 482-0408
Larry Moskovitz

Retail and Wholesale

Plant Specialties: Orchids: miniature, Cattleya, species, Masdevallia, Oncidium. Unusual genera include Zygopetalum, Laelia .

History/description: Larry Moskovitz' studies in photography, botany and ornamental horticulture combined with a lifelong fascination with orchids eventually led him into the nursery business. Opened in 1985, the nursery will board out-of-bloom plants, repot your cymbidium, and give talks to groups. Past president of the San Francisco Orchid Society, Larry gladly shares his knowledge with visitors.

How to get there: Call for directions.

General information: Open by appointment only. They will ship plants. Plant list available for S.A.S.E.

County: Alameda

THE ORCHID RANCH: FORDYCE ORCHIDS, ORCHIDS ORINDA, TONKIN ORCHIDS

1330 Isabel Avenue
Livermore 94550
(510) 447-7171
The Fordyces, Michaels, and Tonkins

Retail and Wholesale

Plant Specialties: Orchids, especially Cattleyas, mostly miniatures (Fordyce), Phalaenopsis (Orinda), Paphiophyllums (Tonkin).

History/description: This collaborative venture started with Mr. Fordyce in 1945. Trained in horticulture and previously employed by a major orchid grower, he soon attracted the help of his family. Then in 1980, the orchid-loving Michaels family joined in for their retirement caper. The Tonkins signed on in 1986. Sounds like more fun than Rossmoor. All three families do their own hybridizing and have many one of a kind specimens, the results of unusual combinations of species. There are display greenhouses.

How to get there: Take the Portola exit from Highway 580 east. Continue to stop light. Right on Murietta for 1 mile. At second light, right on Stanley Boulevard for 1 mile. Left on Isabel.

General information: Open Tueday through Saturday, 10 am - 5 pm; Sunday, 1 pm - 5 pm. They will ship plants. Free plant list is available.

County: Alameda

SAN MIGUEL GREENHOUSES
936 San Miguel Road
Concord 94520
(510) 798-0476
Pamela and James Leaver

Retail and Wholesale

Plant Specialties: Rare, epiphytic Bromeliads, Tillandsia, Aechmea, Vriesea, many species, some hybrid Neoregelia.

History/description: This family business is an extension of Pamela Leaver's bromeliad hobby which first outgrew her home, then outgrew her rented greenhouse and now (since 1987) happily lodges in a 9,000 sq. Fort greenhouse in Concord. Of course, her growing space is much greater than that, since epiphytic bromeliads can grow almost anywhere. Visitors will enjoy the riotous color inside.

How to get there: Take Highway 24 to Highway 680 north. Take Treat Avenue exit. After 12 traffic lights, go left on San Miguel until #934. Go right over canal behind gate to #936.

General information: Open everyday, 1:00 pm until dark. They will ship plant material and Mrs. Leaver loves to give talks about bromeliads.

County: Contra Costa

HORTICULTURAL ATTRACTIONS

Blake Garden

70 Rincon Road, off the Arlington
Kensington 94707
(510) 524-2449

History/description: Formerly the home of Mr. and Mrs. Anson Blake, this eleven acre garden and house is the official residence of the President of the U.C. system. The garden was designed in the mid 1920s by Mrs. Blake's sister, one of the first UC Berkeley Landscape Architecture students. The house was situated to take advantage of the spectacular views and to serve as a windbreak for the garden. A great variety of microclimates makes possible a variety of garden rooms, including a Redwood Canyon, dry Mediterranean garden, Italianate formal garden, and Australian Hollow. Groups can arrange for tours. Garden volunteers are gladly welcomed.

General information: Open 8 - 4:30, Monday - Friday, closed university holidays.

Heather Farm Garden Center

1540 Marchbanks Drive
Walnut Creek 94598
(510) 947-1678

History/description: Heather was a horse and this was her farm, owned by Mr. Marchbanks, a famous breeder of thoroughbreds. Since 1970 it has been a private, non-profit demonstration garden to educate people about what will grow in Zone 14. Special areas include a Drought-resistant Garden, California Native Garden, Children's Garden, Rock Garden, and a Sensory Garden. Starting in March 1993, free tours will be given on the fourth Saturday of the month, 9-10:30 am.

General information: Garden is open during daylight hours everyday. Office and Library open, 9 am - 1 pm.

Plant sales: Plant sales are held in the spring and fall. The facility is rented by plant societies (daffodil, camellia, chrysanthemum, dahlia, iris, african violet) who also organize shows and plant sales.

Regional Parks Botanic Garden

Wildcat Canyon Road at foot of South Park Drive in Tilden Park
Berkeley 94708
(510) 841-8732

History/description: Part of the East Bay Regional Park district, the Botanic Garden is devoted exclusively to California native flora. Not surprisingly, it has been zealously supported by the California Native Plant Society. Tours of the ten acre garden can be arranged for groups of 10 or more. There is a winter lecture series from October to February. 1990 marked the Garden's 50th anniversary.

Plant sales: Plant sale held on the third Saturday of April.

County: Alameda

The Ruth Bancroft Garden

Walnut Creek (415) 824-2919 (tours), (510) 210-9663

History/description: In 1972, with his walnut orchard in decline and her 1000 plants in pots proving too time-consuming to care for, the Bancrofts began this exquisite 2.5 acre garden of cacti, succulents, and companion plants. Creating mounded beds where once walnut trees grew, Ruth Bancroft artfully planted her collection of plants. The first sponsored project of the Garden Conservancy, the garden is being transferred to this organization and is locally supported by a developing Friends group. Friends provide free tours for members, $3 for non-members, and offer special events.

General information: Garden is open from mid-April to mid-October. Reservations required.

County: Contra Costa

UC BOTANICAL GARDEN
University of California
Centennial Drive
Berkeley 94720
(510) 642-3343

History/description: A venerable 101 years old, this 34 acre garden has special collections of California Native Plants, New World Desert Plants, and Asian Plants. Free docent tours, 1:30 pm, Saturday and Sunday. Their membership support group arranges a wide variety of lectures and workshops. They also run the Visitor Center which includes a garden shop selling plants, books, and other garden items.

General information: Garden is open every day except Christmas. Visitor Center open from 10-4 every day.

Plant sales: Their big plant sale is on Mother's Day weekend; smaller sales are held throughout the year.

County: Alameda

Other Attractions:

Berkeley Municipal Rose Garden
Euclid Avenue at Bay View Place
Berkeley 94708
(510) 644-6530

3,000 roses on terraced amphitheater.

Dunsmuir House
2960 Peralta Oaks Court
Oakland 94605
(510) 562-0328

Forty acre park-like garden designed by John McLaren for Dunsmuir and Hellman families' estate.

Lakeside Park Center and Gardens
666 Bellevue
Oakland 94610

Five acre garden; venue for meetings, shows and sales by bonsai, dahlia, orchid, rose, iris, lily and African violet societies.

Markham Nature Park and Arboretum
End of La Vista, off Clayton Road
P. O. Box 12672, Casa Correo Station
Concord 94527

Combination of arboretum, display garden and creekside, natural area on fifteen acres.

Morcum Amphitheatre of Roses
700 Jean Street
Oakland 94610
(510) 658-0731

500 varieties of roses on seven acres.

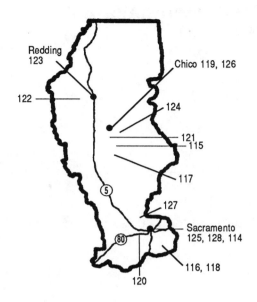

SACRAMENTO VALLEY

**SACRAMENTO
BUTTE
YOLO
SHASTA
SUTTER**

BELL HARDWARE AND NURSERY

5484A Dewey Drive
Fair Oaks 95628
(916) 961-6515
Doug Demetre

Retail and Wholesale

Plant Specialties: Broad selection of Ornamentals - many Ground Covers (30 varieties), Trees, especially Fruit trees, Japanese Maples, Crepe Myrtles, some Perennials.

History/description: Doug Demetre put himself through UC Davis doing landscape work. Aided by a degree in economics and a minor in landscape horticulture, he started his own landscape business in 1962. Already growing plants for his own needs, he bought an existing hardware/nursery operation in 1969. Under his ownership, the nursery side of the business has steadily expanded. Today about 80% of his stock is produced on site. In addition to the Capitol Gardens, he recommends you visit Jensen's Botanical Garden on Fair Oaks Boulevard in north Sacramento and the Rose Garden in McKinley Park.

How to get there: From Highway 80, take Madison exit east. Go north on Dewey. Nursery next to Bel Air market.

General information: Open everyday, 8 am - 6 pm. In Summer, 8 am - 8 pm. Catalogue available for $1.

County: Sacramento

CANYON CREEK NURSERY

3527 Dry Creek Road
Oroville 95965
(916) 533-2166
Susan and John Whittlesey

Retail

Plant Specialties: Perennials. Historical varieties of Dianthus, Sweet Violet. Large selection of Euphorbia, Aster, Campanula, Salvia and Geranium.

History/description: Six years ago Susan and John Whittlesey decided to start their own nursery and growing grounds after having worked for a large mail-order nursery in Washington state. Most of their plants are field-grown and hardy, sold in 3.5" - 5" pots. They are located at 800' elevation in an oak-woodland canyon. Bidwell Park in Chico is nearby.

How to get there: Call for directions.

General information: Open by appointment only Mail order catalogue available for $1.

County: Butte

CORNFLOWER FARMS

P. O. Box 896
Elk Grove 95759
(916) 689-1015
Ann Chandler

Wholesale

Plant Specialties: California native plants, Drought-tolerant plants, California revegetation material.

History/description: This business began because of the demand for quality lining-out and container stock of California native and drought-tolerant plant material. Ten years later the demand has intensified, but, thanks to Cornflower Farms, so has the availability of garden-caliber, water-conserving plants. The U.C. Davis Arboretum is nearby.

How to get there: Call for directions.

General information: Open by appointment only. They ship and deliver plants.

County: Sacramento

COSTA NURSERY

670 Block Road,
P. O. Box 67
Gridley 95948
(916) 846-0160, FAX (916) 846-6641
Bill Shere, Tobias Smith

Retail and Wholesale

Plant Specialties: Fruit and Nut Trees, sold bare root in winter, in containers after April. Their specialty is grafting on semi-dwarf rootstocks. Custom grafting for espalier, horizontal cordons.

History/description: These two longtime friends had trouble finding bare root shade trees for their own purposes and so decided to go into business. Bill Shere, an agricultural consultant, and Tobias Smith, a rice grower, brought real knowledge to their undertaking. They will do custom growing and grafting and offer site-specific consultations.

How to get there: From Highway 99 going north (before the Gridley traffic lights), go left (west) on West Evans Reimer Road for 2 miles, then left on Block Road for a 1/4 mile.

General information: Open everyday, daylight to dark. Closed holidays. Plant list is available.

County: Butte

COTTAGE GARDENS

11314 Randolph Road
Wilton 95693
(916) 687-6134
James McWhirter

Retail and Wholesale

Plant Specialties: Iris - 1,170 tall-bearded and median iris cultivars.

History/description: James McWhirter raised iris as a child growing up in Tennessee. He remains a stalwart collector and enthusiast about his former state's State Flower. Very active in the American Iris Society, he travels as much as possible to collect from other hybridizers. He began his iris business in 1973 in Hayward, moving to Wilton in 1985. All species of iris are on display in his garden although only tall-bearded and medians are sold. This iris judge gladly gives talks and answers questions.

How to get there: From Highway 99 south heading toward Stockton, exit on Dillard. Travel east for 5 miles, then left on Randolph.

General information: Open mid-April to mid-May, anytime during daylight hours. They ship plants. Mail order catalogue available for $1.

County: Sacramento

COVERED BRIDGE GARDENS

1821 Honey Run Road
Chico 95928-8850
(916) 342-6661
Betty and Harry Harwood

Retail

Plant Specialties: Daylilies - 700 registered cultivars.

History/description: Retired dentist Harry Harwood started collecting daylilies in 1981. Betty, the pragmatic partner, figured as long as they were working that hard on their collection, they might as well try to sell some, even if only to finance more plants. Their two acre garden has a 400' creek frontage and is a designated display garden of the American Hemerocallis Society. They introduce new cultivars each year, will contract grow and give slide talks. The nursery is named for an historical covered bridge two miles from their garden.

How to get there: From Highway 99 going north, take Paradise exit. Take Paradise Skyway for 1 mile. Left on Honey Run Road. After 5 miles the road forks at the covered bridge. Go right.

General information: Open June and July, everyday, 9 am - 4 pm. They ship plants. Mail order catalogue available for $1, refundable with purchase.

County: Butte

FLOWERS AND GREENS

P. O. Box 1802
Davis 95617
(916) 756-9238
Roy Sachs, Sue Cartwright

Retail and Wholesale

Plant Specialties: Alstroemeria (rhizomes), Freesia, Perennials. heat-tolerant Buddleias. They also prepare bouquets of cut flowers to sell locally.

History/description: Both owners are professional horticulturists connected with the Environmental Horticulture Department at U.C. Davis. Their work on clonal development of eucalyptus species led to their interest in more colorful plants, hence this nursery's inception in 1988. They both give lectures and suggest you visit the Ruth Risdon Storer Gardens at the UC Davis Arboretum, and the Capitol Park Gardens in Sacramento.

How to get there: Call for directions.

General information: Open by appointment only. Catalogue is available for S.A.S.E. and they will ship your orders.

County: Yolo

GOLD RUN IRIS GARDEN

P. O. Box 648
Durham 95938
(916) 894-6916
Kathy Hutchinson, Sara Skillin

Retail and Wholesale

Plant Specialties: Iris, primarily Tall-bearded, Standard Dwarf-bearded. Some Aril-bred, Louisiana, Spuria, Siberian, Dutch.

History/description: Kathy Hutchinson's iris collection just outgrew her garden space. Requiring more room and more money to finance her growing collection, she went into business in 1989 with her mother. They maintain a display garden and focus their hybridizing efforts on increasing the color range of bearded iris.

How to get there: From Chico, go south on Highway 99, past Highway 149 cutoff, past Nelson Road. Watch for Skillin Lane on your right, marked by "Gold Run" sign. Travelling north on Highway 99, Skillin Lane will be on your left past Gridley, past Cottonwood Road.

General information: Open during bloom season, April, 10 am - 5 pm, or by appointment. Orders taken through July. They ship plants. Catalogue available for $1.

County: Butte

JIM AND IRENE RUSS QUALITY PLANTS

Buell Road, Ono (nursery location)
HCR 1, Box 6450, Igo 96047 (mailing address)
Ono (916) 396-2329
Irene and Jim Russ

Retail

Plant Specialties: Succulents, all hardy including Sedum (80 varieties), Sempervivum (150 varieties), Dudleya, Lewisia. Some Cacti and related plants, Fouquieria fasciculata, Eriocarpus.

History/description: Irene Russ has been an alpine succulent collector for 34 years, joined for many years by her equally enthusiastic husband. They inherited the business from respected grower, Helen Payne, and remain one of the few growers of hardy succulents in California, and the biggest. They have a small display garden and suggest you stop by on your way to the nearby Trinity Alps and Marble mountains.

How to get there: Located west of Redding at 1,500' in the foothills of the Shasta - Trinity National Forest. Call for directions.

General information: Open by appointment almost anytime, call ahead. They will ship plants. Catalogue is available for $1.

County: Shasta

MAXIM'S GREENWOOD GARDENS

2157 Sonoma Street
Redding 96001
(916) 241-0764
Georgia Maxim

Retail

Plant Specialties: Eleven kinds of Iris, including Dwarf, Tall-bearded, Louisiana, Siberian, Spuria, Japanese, Aril-bred (Oncocyclus x Tall-bearded). Daylilies, diploids and tetraploids; nursery has introduced 12, including 'Georgia Maxim'. Also, 350 named varieties of Daffodils.

History/description: Beware of anyone bearing gifts. In 1954 when Georgia Maxim and her husband moved into their new home, a friend gave them a tall-bearded iris as a house-warming present. Her late husband, a Highway Patrol district chief, was instantly enchanted and started getting involved in regional and international iris societies. His collection grew and Georgia started selling surplus plants. Her green touch with sales has produced a thriving business. She is a master judge of the National Iris Society and the State Council of Garden Clubs and maintains a display garden.

How to get there: In Redding, from Highway 5, take Cypress Street exit. Go left to end which is Pine Street. Go right. At light, go left on South Street across the railroad tracks. Go left on Court Street. Right on Sonoma Street up the hill.

General information: Open by appointment only. Best time to see the nursery is March through August. They ship plants. Catalogue is available for mail order.

County: Shasta

MENDON'S NURSERY

5424 Foster Road
Paradise 95969
(916) 877-7341
Jerry Mendon, Joanne and John Mendon

Retail and Wholesale

Plant Specialties: Trees and Shrubs, including unique collections of Japanese Maples, Dwarf Conifers, Dogwood.

History/description: After Cal Poly, the Mendons started a small general nursery in San Gabriel. Fascinated with palms, they also had a professional tree-moving business creating many an instant oasis throughout Los Angeles. Seeking a calmer, more rural environment, they found Paradise and vowed never to open another nursery. They started selling fruits and nuts harvested on their eight acre property, got a few more fruit trees, sold a few to neighbors, and within four years they were back in business. They alert you to the Feather River Hospital Auxiliary's late spring garden tour in Paradise and the garden tour in Chico put on by St. John's Episcopal Church.

How to get there: From Chico take Skyway to Paradise. At traffic light go right for one block. Go right on Foster for 1/2 mile.

General information: Open Monday through Saturday, 8 am - 5 pm, closed major holidays.

County: Butte

MIGHTY MINIS

7318 Sahara Court
Sacramento 95828
(916) 421-7284
Jean Stokes

Retail

Plant Specialties: 800 varieties of miniature African Violets (Saintpaulia) including 'Bustle Back', 'Wasps', longifolia, variegated and tiny ones, mostly registered forms.

History/description: Jean Stokes has been growing African Violets since 1970 when they were first hybridized. An avid collector, she started this business in 1985 and it has grown to become almost a full-time concern. She recommends you visit in January and February for best bloom. She gives talks to groups.

How to get there: From Highway 80 in Sacramento, go east on Florin Road. Go right on Lindale. Sahara Court is off Lindale after the four way stop.

General information: Open by appointment only. Monday through Friday, except in summer. They ship plants all year, unless it gets too cold. Catalogue available for $2.

County: Sacramento Valley

THE PLANT BARN

406 Entler Avenue
Chico 95928
(916) 345-3121
Ilona and David Cronan

Retail

Plant Specialties: Unusual Annuals, Perennials, including Lavandula, Artemisia, Verbena, Phlox and indoor Tropicals. Large selection of 6" potted color.

History/description: Combining his degree in horticulture with her background in art accounts for this profusion of growing color. Started in 1980, The Plant Barn now covers two acres with poly quonset houses, a gift shop, and a brilliant display garden. In December, 10,000 pointsettias make a grand holiday statement. While you are in Chico, check out Bidwell Park (east of Esplanade), Stansbury House (west of Esplanade) and the Bidwell Mansion.

How to get there: From 99 north, left on Entler (next left after a golf course). Nursery is in 1 mile.

General information: Open March - June and December: Monday through Saturday, 9 am - 5 pm. Sunday, 1 pm - 5 pm. They offer tours and talks and will custom grow plants.

County: Butte

RIVER OAKS NURSERY

4827 Pacific Avenue (mailing address)
Pleasant Grove 95668
(916) 655-3591
Bobbi Coggins

Wholesale

Plant Specialties: California native plants, Drought-tolerant plants, specimen Trees.

History/description: Although River Oaks officially opened in 1990, Bobbi Coggins has been a plant broker since 1977. She is a revegetation specialist for waterways.

How to get there: Call for directions.

General information: Open to wholesale trade only by appointment, Monday through Friday, 8 am - 5 pm. Public may visit only with a landscaper.

County: Sutter

RORIS GARDENS

8195 Bradshaw Road
Sacramento 95829
(916) 689-7460, FAX (916) 689-5516
Joe Grant, Manager

Retail

Plant Specialties: Over 350 varieties of Tall-bearded Iris, Japanese Iris (starting 1993)

History/description: Postcard perfect. Picture the old farmhouse next to a creek in the shade of mature hardwood trees, with wild peacocks, geese, ducks and egrets feeding nearby. Then surround it all with fifteen spectacular acres of irises in every color of the rainbow. All this just fourteen miles from downtown Sacramento. Roris Gardens gives talks to groups by arrangement.

How to get there: From downtown Sacramento, take Highway 50 east. Take Bradshaw Road exit, go south on Bradshaw for 8 miles.

General information: Open during their Iris Festival from mid-April to mid-May, 8 am - 5 pm. They ship plants anywhere. Color mail order catalogue (72 pages) available for $3, refundable with purchase.

County: Sacramento

HORTICULTURAL ATTRACTIONS

Capitol Park
1300 L Street
Sacramento
(916) 445-3658

History/description: These forty acres which surround the capitol buildings, contain perhaps the largest and most extensive collection of mature trees in California. The historic insectary in the Park's center provides visitors with self-guiding maps to 341 special trees; there are 200 more which do not even make the map. Planting began in 1870 and continues in a sporadic fashion. The azalea collection in bloom has been described as "socko."

Davis Arboretum
Temporary Building 32, La Rue Road
Davis 95616
(916) 752-2498 - Friends of the Davis Arboretum

History/description: Davis Arboretum has 150 developed acres emphasizing drought-tolerant, low-maintenance plants. There is no admission charge although parking costs $2. Special gardens include an Oak Grove, Native Plant Garden, White Flower Garden and dry land Demonstration Garden. The extensive Eucalyptus and Acacia collections, damaged during the 1989 freeze, are being rebuilt. The Friends of the Davis Arboretum, a membership support group, organizes monthly lectures, Sunday tours and workshops during the academic year.

General information: Open everyday.

Plant sales: Big fall plant sale. Monthly sales for Friends only, during academic year.

Other Attractions:

Bidwell Park
Chico
(916) 895-4972

Ten miles-long urban oasis with many old, stately trees.

Genetic Resource Center
US Forest Center
Chico
(916) 895-1176

Nature trail with many labelled examples of Center's historic plant introductions.

C. M. Goethe Arboretum
California State University - Biology Department
Sacramento 95819-6077
(916) 278-6077

Six acre collection of trees and shrubs.

Charles Jensen Botanic Garden
8520 Fair Oaks Boulevard
Carmichael
(916) 944-2025

Temperate, moist woodland plants along a creek.

Rose Garden in McKinley Park
H and 33rd Streets
Sacramento 95816
(916) 277-6060

Roses and companion plants in park setting.

Shepherd Garden and Art Center
3330 McKinley Boulevard, next to McKinley Park
Sacramento 95816
(916) 443-9413

Site of about fifty garden and art groups' meetings, shows and sales. Big plant sale in March.

William Land Park
Sutterville Road and Land Park Road
Sacramento
(916) 277-6060

Old rock garden and new color garden designed by Daisy Mah.

132 Where On Earth!

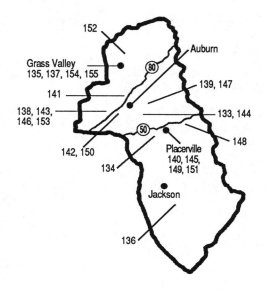

FOOTHILLS

**EL DORADO
NEVADA
CALAVERAS
PLACER**

BLUE OAK NURSERY

2731 Mountain Oak Lane
Rescue 95672
(916) 677-2111
Eileen and Vanessa Beegle

Retail and Wholesale

Plant Specialties: California native plants, Drought-tolerant plants, Oaks and selected Ornamentals. Introduced Rancho Santa Ana Botanic Garden's Penstemon 'Midnight' and P. 'Firebird'.

History/description: This mother-daughter team has been operating Blue Oak Nursery in the oak-woodland area of the foothills since 1972. Eileen Beegle saw the need for supplying appropriate native plants for eager gardeners and landscape designers in the area. Their scenic location and specialty plants now beckon visitors from afar. They contract grow for revegetation projects.

How to get there: Take Highway 50 to Cameron Park Drive exit. Follow Cameron Park Drive (its name changes to Starbuck after crossing Green Valley Road) to Deer Valley Road. Go right on Deer Valley to Mountain Oak Lane.

General information: Open Monday through Thursday, 9 am - 5 pm; Saturday, by appointment only. They deliver within a 50 mile radius and ship mail order plants anywhere. Catalogue available for $1. Slide talks arranged for interested groups.

County: El Dorado

BLUEBIRD HAVEN IRIS GARDENS

6940 Fairplay Road
Somerset 95684
(209) 245-5017
Mary and John Hess

Retail

Plant Specialties: Iris — tall-bearded, border-bearded, dwarf-bearded.

History/description: This garden nursery has been in its present location since 1988, although it previously existed, primarily as a bird sanctuary, for nine years elsewhere. This garden location is being developed on a Victorian theme with gazebos, benches, archways and costumed employees reflecting a turn-of-the-century civility. Landscaped walkways through three miles of planted iris rows should be completed by 1993.

How to get there: From Highway 50 just west of Placerville, take the Missouri Flat exit. Turn south on Missouri Flat Road to Pleasant Valley Road. Turn left on Pleasant Valley Road to E16 (about 12 miles). Turn right on E16 (also known as Mt. Aukum Road). Turn left on Fairplay Road for 1/4 mile.

General information: Open during bloom season (third weekend in April - fourth weekend in May), Tuesday - Sunday 10 am - 5 pm. Rest of year by appointment. Will ship anywhere in USA. General catalogue available for $1. Antique list (pre-1964) also available for $1. Please specify which catalogue you want.

County: El Dorado

BRINSLEY'S ORCHIDS

14528 Lime Kiln Road
Grass Valley 95949
(916) 268-1509
Dorothy Mottweiler

Retail and Wholesale

Plant Specialties: Orchids, 150 varieties, emphasizing miniature Cattleyas, Phalaenopsis, standard and miniature Cymbidiums, Exotics.

History/description: A lifelong orchid hobbyist, Dorothy Mottweiler retired from nursing in 1978 and decided to add an orchid greenhouse to her home to serve as its solar collector. The effect on her utilities bill is unknown, but the extra space begat more orchids and, hence, a business. She now tends a 500 square foot greenhouse stacked to the gunwales with orchids, including many exotics collected on trips to Peru and Thailand. She gives seminars in the Foothills area.

How to get there: Call for directions.

General information: Open by appointment only. Plant catalogue is available for $1 and S.A.S.E.

County: Nevada

CALAVERAS NURSERY

1622 Highway 12
Valley Springs 95252
(209) 772-1823
Mick and Pat Stoppard

Wholesale and Retail

Plant Specialties: Specialty Oaks (27 varieties) from all over the world, also Redbud, Toyon.

History/description: Mick Stoppard helped finance his way through college by working at Boething Treeland. Armed with degrees in zoology and plant science, he launched Calaveras Nursery in 1977. Worth a visit for its gardenlike setting in the foothills (700' elevation), this ten acre nursery is nestled in a wooded area with a fine creek. Worth a visit as well is Grinding Rock State Park, just north of Jackson on Highway 88 and Micke Grove Park in Stockton.

How to get there: From Highway 99, go east on Highway 12. Nursery is two miles past the town of Valley Springs, approximately thirty-five miles south of Sacramento and eleven miles north of Stockton.

General information: Open by appointment only. For wholesale customers, 8 am - 5 pm, Monday through Friday. A plant list is available.

County: Calaveras

FOOTHILL COTTAGE GARDENS

13925 Sontag Road
Grass Valley 95945
(916) 272-4362
Carolyn Singer

Retail

Plant Specialties: Deer, drought and cold (to 10 degrees) resistant Perennials, including Penstemon, Artemisia, Achillea, Santolina and Lavandula. Mediterranean plants, Rock Garden plants/Alpines, Thyme (15 different varieties)

History/description: This longtime gardener who had worked the soil in Sebastopol, Colorado and Montana, moved to Grass Valley for solitude and for horticultural therapy. One thing led to another. Her love of the land soon had her sharing her knowledge. She taught classes; students wanted her plants. And so this nursery started in 1977 on the site of the old Sonntag homestead. There is a display garden. Recommended places to visit while in the area include Empire Mine and the old rose gardens of St. Joseph's Museum in Grass Valley. She alerts you to the interesting meetings, shows and sales of the Perennial Plant Association of the Foothills and Valley.

How to get there: Call for directions.

General information: Open Tuesday, Thursday, Saturday, 9 am - 2 pm., and by appointment. They will ship plants March to May and September to November. Very detailed catalogue is available for $3 which includes drawings, suggested planting combinations and hardiness ratings.

County: Nevada

FOWLER NURSERIES, INC.

525 Fowler Nurseries, Inc.
Newcastle 95658
(916) 645-8191
Robert Fowler; Richard Fowler, Mgr. Retail Section

Retail and Wholesale

Plant Specialties: Fruit and Nut Trees, old varieties and specially tested new introductions. Seasonal Bare Root stock, including Ornamental Trees, Grapes, Berries, etc. Ornamental Trees, Shrubs, and Perennials stocked from other growers.

History/description: This multi-faceted family business was started 80 years ago by grandfather Eugene Fowler, whose pioneering predecessors settled in Placer County back in the days when it grew most of the fruits and nuts for the whole state. All generations have worked in the business. Important lessons learned at home have been supplemented by training at UC Davis and Cal Poly San Luis Obispo. Now daughter Nancy oversees the commercial operation dealing with orchardists, son Richard manages the retail sales, and son-in-law Everett Johnson raises forest seedlings for reforestation projects and Christmas Tree farmers.

How to get there: From Highway 80, take Sierra College Boulevard exit. Go north for 6.5 miles. Right on Highway 193. After 1 mile, go left on Fowler Road for 1/4 mile.

General information: Open Monday through Saturday, 8 am - 5 pm; Sunday, 9:30 - 4 pm. They will ship mail orders (bare root only). Retail plant list available for $4. Wholesale fruit and nut plant list $3. They produce a newsletter, give pruning workshops and orchard clinics, and can do contract growing.

County: Placer

Foothills

FOX RUN NURSERY
P. O. Box 711
Georgetown 95634
No phone
Darby Haines

Retail and Wholesale

Plant Specialties: Cold hardy and heat tolerant Perennials, 100 varieties of culinary and medicinal Herbs.

History/description: This former resident of northern Scotland combines her love of English garden plants with the realities of gardening in the California foothills. She grows and tests 100 species of perennials for their tolerance to cold and heat at the nursery (2600' elevation). A display garden is being established. Fox Run will custom grow plants for customers.

How to get there: 2.5 miles east of Georgetown on Wentworth Springs Road (continuation of Main Street). For the Open Houses, signs will be posted on Wentworth Springs Road.

General information: Open by written appointment only (please include your telephone number so that directions can be given over the phone). The nursery has Open Houses on weekends in May and June and sometime in the fall.

County: El Dorado

GOLD HILL NURSERY
6000 Gold Hill Road
Placerville 95667
(916) 622-2190
Al Veerkamp

Retail and Wholesale

Plant Specialties: Japanese Maples, Dogwoods, unusual Trees and Shrubs, including Chinese Magnolias, weeping forms of Beech, Gingko, "oddball stuff for hot summers and low temperatures".

History/description: Born and raised in Placerville, Al Veerkamp worked for the Department of Agriculture following his graduation from UC Davis. In 1957 there was no nursery in El Dorado County so he started growing general nursery stock. Over a period of time, he started to specialize in unusual grafted conifers and deciduous trees. His display garden, a miniature arboretum, offers a feast for the connoisseur.

How to get there: West of Placerville on Highway 50, take the Placerville Drive exit. Go north on Placerville Drive half mile past Raleys grocery store. Left on Pieroz Road. Left at stop sign on Cold Springs Road. Continue 7 miles to Gold Hill Road.

General information: Open Wednesday through Sunday, 9 am - 5 pm.

County: El Dorado

GREEN LEAF IN DROUGHT TIME

P. O. Box 597
Meadow Vista 95722
(916) 878-2063
Beth Canfield

Retail and Wholesale

Plant Specialties: California native plants - Shrubs and Ground Covers, especially those well-suited to the Foothills.

History/description: Another wonderful instance of a hobby gone beserk. Beth Canfield started out taking some agricultural classes at Sierra College, got interested in native plants, got interested in the California Native Plant Society, got interested in growing plants for CNPS sales...now she has 1/4 acre under propagation.

How to get there: Call for directions. Leave request on machine.

General information: Seasonally open, Spring and Fall, call first. Plant list is available.

County: Placer

HIGH RANCH NURSERY

3778 Delmar Avenue
Loomis 95650
(916) 652-9261
John C. Nitta

Wholesale

Plant Specialties: General woody ornamentals. California natives - Oaks (12 varieties). Hardy and drought-tolerant Perennials. Lagerstroemia 'Seminole', L. 'Cherokee' and L. indica (mildew resistant form of crape myrtle) 'Muskogee', 'Tuscarora', 'Natchez'.

History/description: Educated at UC Davis, John Nitta started High Ranch Nursery in 1976 as a shade tree nursery. Increased public interest and general unavailability of choice plants shifted his focus slightly to include California natives. He has a display garden and can sometimes be persuaded to do contract growing.

How to get there: Call for directions.

General information: Open to wholesale trade only, Monday through Friday, 7 am - 5 pm. Retail customers welcome only with a landscape professional. Plant list available for S.A.S.E.

County: Placer

Foothills

LAKE'S NURSERY
8435 Crater Hill Road
Newcastle 95658-9666
(916) 885-1027
David and Judy Lake

Retail

Plant Specialties: Japanese Maples, Bonsai starters, Trees and Shrubs.

History/description: When the seedlings for his organic vegetable garden and his growing collection of garden plants overran all available space, David Lake realized he needed to start selling. And the public wanted to buy as there were no nurseries in the area in 1977. A former firefighter, he and his wife have worked full-time in the nursery since 1983. The 2.5 acre location has a Japanese display garden with a 22,000 gallon koi pond.

How to get there: From Highway 80 east, take Highway 193 exit to Lincoln. Curve around back under road through stop sign to Ophir Rd. Left on Lozanos for one mile to stop sign. Go left and right around a store and up the hill. At first stop past a school, go left on Wise Rd. Near fire station at top of hill, go left on Crater Hill Rd.

General information: Open Monday through Saturday, 8 am - 5:30 pm; closed Sunday. Customers are invited to bring a drawing of their garden site and the Lakes will help develop a planting plan.

County: Placer

LOTUS VALLEY NURSERY & GARDENS

5606 Peterson Lane,
P. O. Box 859
Lotus 95651
(916) 626-7021
Bob Davenport, Joe House

Retail

Plant Specialties: Ornamental grasses, Drought-tolerant plants, Perennials, Shrubs and Trees; featuring many California native plants.

History/description: These two delightfully compulsive gardeners started selling just a few plants from their own rich and varied garden. Word obviously got out. Now they are full-time plant producers, operating in a four acre valley near the south fork of the American River with a beautiful view of the mountains. Their rock garden was created from granite blasted to prepare a building site for the original (now burned) Victorian ranch house. Their garden grows like topsy, using Mediterranean plants in all sorts of unusual combinations. Tours may be arranged and they give talks and slide presentations. They suggest you combine your visit with a trip to the nearby the Marshall Gold Discovery State Park.

How to get there: Hwy. 50 East to Shingle Springs. Follow signs to Lotus. Turn left on Bassi Rd. Right on Peterson Lane. From Auburn: Take Hwy. 49 to Lotus. Right on Lotus Road Right on Bassi Road Right on Peterson Lane.

General information: Open Wednesday through Sunday, 9 am - 5 pm. Deliveries to Placerville/Sacramento area. Catalogue available for $2.

County: El Dorado

MAPLE LEAF NURSERY

4236 Greenstone Road
Placerville 95667
(916) 626-8371
Robert Barnard

Retail

Plant Specialties: Drought-tolerant plants, deciduous Azaleas, interesting Trees, Shrubs and Perennials. Special collections of Corylopsis, Hamemelis, Cornus, including new hybrid (Kousa x Florida).

History/description: Longtime nurseryman, Bob Barnard, started Maple Leaf in 1987 for the sole purpose of introducing new and different plants to the public. The nursery, located on an east facing oak-studded savanna, consists of a small garden and a half acre of nursery stock.

How to get there: Nursery is located half mile south of Highway 50 on the east side of Greenstone Road, between the towns of Shingle Springs and Placerville.

General information: Open by appointment only, Monday through Thursday. Catalogue available for $1.50. They ship plants.

County: El Dorado

MATSUDA LANDSCAPE AND NURSERY

4888 Virginiatown Road
Newcastle 95658
(916) 645-1820
Hiroshi Matsuda

Retail

Plant Specialties: Bonsai stock (2" pots to 15 gallon cans), including Maples and Pinus thunbergii shaped to many sizes.

History/description: As a designer and builder of Japanese gardens with a degree in Landscape Architecture from Cal Poly Pomona, Hiroshi Matsuda was well aware of the need for choice plants. With the opening of the nursery in 1991, Matsuda's has a monopoly on garden-making. He specializes in custom-designed Sai Kei or miniature landscapes which use planted bonsai. The nursery contains many such landscapes and other bonsai displays in its shade houses.

How to get there: Take Highway 80 to Sierra College Boulevard. Left on Sierra College Boulevard. Right on Highway 193. Left on Fowler Road. In 1/2 mile go right on Virginiatown Road. Nursery in 0.2 mile.

General information: Open Thursday through Saturday, (9:30 am - 5:00 pm) or by appointment. Hiroshi Matsuda gives Bonsai classes at the nursery.

County: Placer

REDBUD FARMS

6300 Highway 193
Georgetown 95634
(916) 333-2300
Martha and David Cox

Retail and Wholesale

Plant Specialties: Redbuds, Herbs, Perennials, Roses, Drought-tolerant and Deer-resistant plants, Fruit, Shade and Nut Trees. Garden supplies.

History/description: The growing ground nursery of these xeriscape specialists is at 3000' elevation beyond the PG&E lines in the Sierra Nevada foothills. Since 1985 they have operated their business and lives completely on alternate energy systems. And they still say it's fun. No traffic jams. Abundant wildflower walks. And the fishing is great. They'll tell you where. Their retail nursery is in town.

How to get there: Take Highway 80 to Auburn, then Highway 49 south to Cool. Left on Highway 193 to Georgetown. Look for sign on right. Or take Highway 50 to Placerville. Highway 49 north to Cool and follow directions above.

General information: Retail open Monday, Tuesday, Thursday, Friday, Saturday 10 am - 5 pm; Sunday 10 am - 3 pm. Wholesale - by appointment only. Closed December 25- January 6 and for a bit in August. They will make deliveries to the Georgetown divide and neighboring counties. They produce a Monthly Specials list and newsletter and give talks to groups.

County: El Dorado

ROSE ACRES

6641 Crystal Blvd.
Diamond Springs 95619
(916) 626-1722
Muriel and Bill Humenick

Retail

Plant Specialties: Roses, especially the single (five-petalled) varieties and the Hybrid Musks.

History/description: The Humenicks have been busy judging and lecturing on the subject of roses since 1957. In 1980 they started this nursery to supply customers with unique and unusual cultivars. They now have 5,000 roses, including many species and old-fashioned types, on their four acre site. Attempting semi-retirement, the Humenicks have made arrangements with Roses and Wine in Shingle Springs to ship their plants. They still host their annual Rose and Wine tour on the Saturday before Mother's Day. And there are several interesting cemeteries with unusual, old plantings nearby.

How to get there: Take Highway 50 east from Sacramento for about 30 miles to El Dorado Road exit. Right on El Dorado Road for 2 miles until it dead-ends. Left onto Pleasant Valley Road. Right at stop sign onto Highway 49. Continue south on Highway 49 for 5 miles. Right onto Crystal Boulevard.

General information: Open by appointment only. Best viewing time is mid-April through June. Plants available for pick-up on site. They will give slide talks to interested groups. Plant list available for S.A.S.E.

County: El Dorado

ROSES AND WINE

6260 Fernwood Drive
Shingle Springs 95682
(916) 677-9722, FAX (916) 676-4560
Barbara and Wayne Procissi

Retail

Plant Specialties: Heritage Roses, especially repeat bloomers, trailing and semi-procumbent ground cover roses.

History/description: Landscape Architect and licensed Landscape Contractor, Barbara Procissi started this home nursery in 1992, once she and her husband were able to keep the deer from munching on their eleven acre site. Despite the foraging of wild turkeys, a vineyard is also being developed. They give tours of their ten acre domain in the heart of the Mother Lode country and can help design your rose garden.

How to get there: Take Highway 50 east from Sacramento for about 20 miles to Ponderosa exit. Go south on South Shingle Road for 5.2 miles. Turn right on Fernwood. Go 1/4 mile and turn left on Thunder (sign in trees). This is a dirt road. After a bit, it forks; veer right on Magpie. Nursery site at 6271 Magpie.

General information: Open by appointment only. Bare-root plants shipped January and February. They also ship plants for Rose Acres. Mail-order catalogue available for S.A.S.E.

County: El Dorado

SIERRA FOOTHILL MINIATURE ROSES

3778 Delmar Avenue
Loomis 95650
(916) 652-9261, (916) 989-1817
Scott K. Nitta

Retail and Wholesale. Plants sold through High Ranch Nursery at same address, (916) 652-9261 or 989-1817.

Plant Specialties: Miniature roses

History/description: For almost three years Scott Nitta has been making the public aware of field-grown miniature roses. He presently has them for sale in containers and in 8" hanging baskets.

How to get there: Take Highway 80 to Sierra College Boulevard north. Turn left on King Road. Turn left on Del Mar Avenue.

General information: Summer hours - 7 am - 6 pm, Monday to Friday; 8 am - 6 pm Saturday; 11 am - 6 pm, Sunday. Winter hours - nursery closes at 5 pm. They deliver to the greater Sacramento area.

County: Placer

SPOTTED FLOWER NURSERY

4100 Creekside
Shingle Springs 95682
(916) 677-6987
Deb Edwards; Jeannie Ross, Carol Patton, Managers

Retail

Plant Specialties: California native plants and Perennials suitable for valley and foothills. Aquatic plants, unusual Trees and Shrubs.

History/description: Deb Edwards started this business in 1986 because of her obsession with plants and the scarcity of available perennials really suited to the area. New partners and a new location in the heart of downtown Shingle Springs (1991) give a boost to her efforts. They also manage Spotted Flower Market in El Dorado Hills, purveyors of dried flowers and garden accessories. They offer landscape consultation services.

How to get there: Take Highway 50 for 20 miles east of Sacramento to Shingle Springs exit. Continue for 1/3 mile. Immediately after railroad tracks, go right on French Creek Road. Nursery is on corner of French Creek and Creekside Roads.

General information: Open Wednesday through Saturday, 9 am - 5 pm. Sunday, 10 am - 4 pm. Plant list is available. They give slide talks to interested groups.

County: El Dorado

SWEETLAND FARM WHOLESALE NURSERY

27443 Sweetland Road
N. San Juan 95960
(916) 292-3141, FAX (916) 292-9255
Dolores Sommer; Lyle Hatch, Propagator

Wholesale

Plant Specialties: California native plants, Perennials, Cold-hardy, Drought-tolerant plants. All plants suited to the high country, including Jeffrey and Lodgepole pines, plants for the Nevada desert, and riparian plants for revegetation.

History/description: Dolores Sommer started this business in 1985 because of her great interest in California native plants. Located at 2100' elevation in an oak-pine forest valley, Sweetland Farm successfully produces plants appropriate to high elevations in California. Dolores offers plant consultations by phone. Nearby points of interest include: Malakoff Diggins State Park, the Annual Highway 49 wildflower display, and Independence Trail which runs from Nevada City to North San Juan (wheelchair accessible).

How to get there: From Highway 80 take Highway 49/20 north to the northside of Nevada City. On the northside of Nevada City, take Highway 49 (Downievale Highway). Continue for fifteen miles to Peterson's Corner Restaurant. Take next left onto access road, then quickly right on Sweetland Road. Nursery is in 3/4 mile.

General information: Open to wholesale customers, by appointment only. Public may visit with a landscape professional. They will deliver throughout the northern California foothills and valley. They have a wholesale catalogue and will give talks on revegetation to groups.

County: Nevada

Foothills

WAPUMNE NATIVE PLANT NURSERY CO.
3807 Mt. Pleasant Road
Lincoln 95648
(916) 645-9737
Everett Butts

Retail and Wholesale

Plant Specialties: California native plants, especially those appropriate to the Sierra foothills and Sacramento Valley.

History/description: Everett Butts has been committed since 1983 to the selection, propagation and introduction of native species. He has made it his mission to help each resident restore and repair his or her piece of California. Populus fremontii 'Nimbus' was named by him and Zauschneria californica ssp. latifolia "Everett's Choice" was named for him. We're proud of this local boy who grew up loving to play on the greens in Golden Gate Park.

How to get there: From Highway 80, take Sierra College Boulevard which dead ends at Highway 193. Go east to Fowler Road. Go right on Fowler Road to end. Go left on Fruitvale, then quick right on Garden Bar Road. Go about two miles to arterial stop at Mt. Pleasant Road. Proceed straight ahead to gate on right.

General information: Open by appointment only. Inquire about local only delivery.

County: Placer

WEISS BROTHERS NURSERY

10120 Joerschke Drive
Grass Valley 95945
(916) 273-5814
Marty and Dwight Weiss, Growing Ground; Emil Baldoni, Retail Nursery

Retail and Wholesale

Plant Specialties: Ornamental Trees, Shrubs, Vegetables, Annuals. Over 300 varieties (50 genera) Perennials, all tested for heat and cold tolerance.

History/description: The nursery business has made important contributions to the education of our youth, or more exactly to our youth while being educated. Emil Baldoni worked in a nursery during high school while the Weiss brothers supplemented their schoolyear incomes by raising seedlings for the forest service and growing Christmas trees in boxes for sale to grocery chains. Horticulturally educated, these entrepreneurs became landscapers and opened the nursery (1973) to supplement the available supply of plants.

How to get there: Take Highway 80 to Highway 49 north to Grass Valley. Take Brunswick exit, go left on overpass. Take next left. Nursery is visible down slight incline next to highway.

General information: Open Monday through Saturday, 8:30 - 5 pm; Sunday 10 am - 5 pm. Wholesale, by appointment only. They will ship plants. 19 page catalogue is available for mail orders.

County: Nevada

YOU BET FARMS

15595 You Bet Road,
15564 You Bet Road (mailing address)
Grass Valley 95945
(916) 273-5931, (916) 273-1423 (evenings)
Rodger L. Rollings

Retail

Plant Specialties: Cold-hardy Perennials, California native plants, Alpines, Drought-, Deer-, and Fire-resistant plants. Of special interest are Penstemon, hardy Geranium, Achillea, Artemisia and Campanula collections.

History/description: Coming from a family of gardeners and florists, Rodger Rollings has had a lifelong familiarity with plants. In his display garden, he now has hundreds of species of rare and unusual perennials which cater to the climate of the Sierra Foothills. Everything here is organically grown and he will gladly show you how to prepare soil with composted material. Nearby points of interest include Empire Mine State Park and Independence Trail.

How to get there: Take Highway 80 east to Colfax. Take Grass Valley exit. Continue for 8 miles. Right on You Bet Road. Nursery on right in a mile and a half. NOTE: nursery will change its name to Your Bet Gardens and move location in late 1993. Call for new directions.

General information: Open Wednesday through Sunday, 10 am - 5 pm. Free catalogue available upon request. Rodger will give talks to groups about gardening in the foothills.

County: Nevada

HORTICULTURAL ATTRACTIONS

Bourne Mansion

Empire Mine State Park
Empire exit off Highway 49
Grass Valley
(916) 273-8522

Another Willis Polk-designed home for mining magnate Willliam Bourne (see Filoli, San Francisco Peninsula region). A thirteen acre European-influenced garden in the ponderosa forest.

Independence Trail

North of Nevada City on Highway 49
P. O. Box 1026, Nevada City 95959
(916) 272-3823

Five mile, wheelchair accessible wilderness trail on old canal bed.

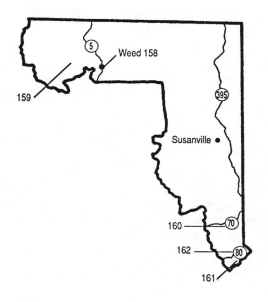

MOUNTAIN REGION

CASCADES
SIERRAS
SISKIYOUS

MENZIES NATIVE NURSERY
P. O. Box 9
Weed 96094
(916) 938-4858
Robert Menzies

Retail and Wholesale

Plant Specialties: Riparian habitat plants, 208 species of montane California native plants.

History/description: Impeccably credentialed horticulturist, this Robert Menzies is grandson of Robert Menzies (founder of Save the Redwoods League), cousin of Arthur Menzies (of Strybing Arboretum fame) and distant relative of Archibald Menzies (pioneer plant explorer). With a background in viticulture at Davis, he started this nursery in Mill Valley in the 1970s and moved to Weed in 1980.

How to get there: Call for directions.

General information: By appointment only. They have a plant list and will ship your orders. They do site-specific consultations, can offer design assistance and will do contract growing. Tours, lectures and workshops are also available.

County: Siskiyou

SCOTT VALLEY PERENNIALS

233 Main Street, or P. O. Box 160
Etna 96027
(916) 467-3764
Barbara Coatney

Retail and Wholesale

Plant Specialties: Sempervirens, Sedums, Alpines and hardy plants, Dianthus, Geraniums, Phlox, Campanulas

History/description: This nursery is the direct result of plant hobbyist Barbara Coatney's frustration with her local supply of interesting plants. A horticultural activist, she solved her problem by opening a nursery in 1981. It now includes a display garden and offers contract growing services. While in Siskiyou County she suggests a visit to the garden of Jeanette Axton in Etna (open Sunday and Monday in summer) and to Mt. Etna summit, Scott Mountain, and Kangaroo Lake for best wildflower viewing.

How to get there: Call for directions.

General information: Call for open hours. They will ship plants and have a plant list available for S.A.S.E.

County: Siskiyou

SIERRA VALLEY NURSERY

P. O. Box 79
Beckworth 96129
(916) 832-0114
Gary Romano

Wholesale

Plant Specialties: California native Trees and Shrubs exclusively, especially those suited to high elevations. Ceanothus cuneatus, C. velutinus, C. prostratus, Arctostaphylos patula, A. nevadensis, Prunus andersonii, Chrysothamnus, Purshia tridentata, Cercocarpus, all Sierra conifers

History/description: There was never really much doubt about what Gary Romano would do. His parents were wholesale cut flower growers in Morgan Hill so quite naturally he studied ornamental horticulture at Cal Poly San Luis Obispo. While doing revegetation work for the Truckee Parks Department and finding so few necessary plants available, he launched the nursery. He also operates a landscape business called Lake Level Environments and can offer site specific consultations. Although Sierra Valley is primarily a contract grower for nurseries and government agencies, they have a large plant selection on site.

How to get there: Call for directions.

General information: Open by appointment; public may visit only with a landscape professional.

County: Plumas

TAHOE TREE COMPANY

401 West Lake Tahoe Boulevard, or P. O. Box 5325
Tahoe City 96145
(916) 583-3911
Leslie and John Hyche, Karen McBride, Fil Aguirre

Retail and Wholesale

Plant Specialties: California native plants, ornamental Trees and Shrubs, cold-hardy Perennials.

History/description: David and Katherine McBride started this nursery in 1954 as an offshoot of his tree service business, her interest in perennials and California native plants, and their combined desire to supply local gardens with healthy plants suited to the area. The second generation is now in control. Daughter Leslie with her background in Landscape Architecture from UC Davis has managed the nursery since 1971. John supervises the 3-acre growing grounds. Other daughter Karen manages the business and retail shop, located in a rustic and airy log building designed by husband, Fil. The 12 acre property includes a garden restaurant. They recommend Julie Carville's *Lingering in Tahoe's Wild Gardens* as a guide to nearby wildflower walks.

How to get there: Located a quarter mile south of Tahoe City on corner of Highway 89 and Granlibakken Road.

General information: Open May through September, 8 am - 6 pm, October through April, 10 am - 5 pm.

County: Placer

THE VILLAGER NURSERY

10994 Donner Pass Road, or P. O. Box 1273
Truckee 96160
(916) 587-0771
Sarah Trebilcock; Eric Larusson, Manager

Retail and Wholesale

Plant Specialties: California native plants suited to mountainous region, Perennials, cold-hardy plants.

History/description: Sarah Ross, a UC Davis educated botanist and ecologist, started the nursery in 1975. She bought a florist shop which came with a good piece of property in the rear, and the nursery was born. In 1980 she obtained her contractor's license and opened an accompanying landscape business. The nursery has a display garden, produces a newsletter about high Sierra gardening, offers design services and will give informal talks to interested groups. Sarah suggests you explore Truckee Regional Park trails which include alpine, high desert, Indian-use plants, wildflower and riparian areas.

How to get there: From the train depot in downtown Truckee, go west under freeway on Donner Pass Road (Main Street). Nursery is on top of hill behind Old Gateway Center. Approach from the rear.

General information: Open Summer, 8 am - 6 pm. Fall and Spring, open according to the weather, call first. Winter (January and February) closed. The florist shop on the premises is open all year (closed Sundays).

County: Nevada

CENTRAL COAST

**SANTA CRUZ
MONTEREY
SAN LUIS OBISPO
SAN BENITO
SANTA BARBARA**

ABBEY GARDEN CACTI & SUCCULENTS

4620 Carpinteria Avenue
Carpinteria 93013
(805) 684-5112, FAX (805) 684-8235
Lem and Pat Higgs

Retail and Wholesale

Plant Specialties: Cacti and Succulents. Most plants are in 2" - 6" pots.

History/description: This business was started in 1967 in Reseda by Charles Glass and Bob Foster to supply collectors of cacti and succulents. Now a 10,000 foot greenhouse and sales area offers over 2,000 species for sale. Potted plants of a single species are grouped together in square wooden flats with their stock plant placed behind them for reference. The Higgs' view their nursery as a zoo for plants, a place to promote the continuation of endangered species since more and more natural habitats are being lost to agriculture. Nearby horticultural attractions include Lotusland in Montecito (due to open Fall, 1993).

How to get there: From Highway 101, take Reynolds Avenue exit. Nursery is adjacent to freeway.

General information: Open Tuesday through Sunday, 9:00 am - 5:00 pm. This owner-operated business ships worldwide. Plant catalogue is available for $2.

County: Santa Barbara

ANTONELLI BROTHERS BEGONIA GARDENS

2545 Capitola Road
Santa Cruz 95062
(408) 475-5222
Skip Antonelli, Linda Bobbitt

Retail and Wholesale

Plant Specialties: Tuberous Begonias (100 varieties), also Fibrous Begonias (50 species), Rex, and Rieger Begonias. Fuchsias (300 cultivars, varieties and subspecies). Shade plants such as Ferns, Mandevilla, Clerodendrum, Cineraria.

History/description: A family business since 1935 when it was started by the present owner's father and uncles, Antonelli's has many begonia hybrids attributed to itself. The current generation, horticulturally educated and trained since birth in the nursery, continues to develop new varieties. Three acres of glass and lath houses provide an enormous selection for their largely retail customer. They alert you to two special plant sales - one in the fall at UCSC Arboretum and the other on Mother's Day at Cabrillo College.

How to get there: In Santa Cruz, travelling south on Highway 1, go right on 41st Avenue South. Go right on Capitola Road. Nursery is on the right.

General information: Open everyday, 9 am - 5 pm. They will ship plants. Catalogue available upon request.

County: Santa Cruz

BAY LAUREL NURSERY

P. O. Box 66595
Scotts Valley 95066
(408) 438-3999
Peter Moerdyke

Retail ($250 minimum order) and Wholesale

Plant Specialties: 100 varieties uncommon Rhododendrons, especially Rhododendron occidentale, Exbury and Kurume azaleas. Trees and Shrubs, special collections of deciduous Magnolia (20 varieties), Gingko, Dogwood, Daphne, Pieris.

History/description: Peter Moerdyke has always been a plant lover. A Cal Poly graduate, he bought out Boulevard Garden Rhododendron Nursery in Palo Alto in 1977 and moved it to Scotts Valley. A display garden is in the works. He suggests you combine your visit with a trip to Henry Cowell Park on the UCSC campus.

How to get there: Call for directions.

General information: Open by appointment only. They produce a free catalogue and will deliver to the immediate area.

County: Santa Cruz

BIO-QUEST INTERNATIONAL

P. O. Box 5752
Santa Barbara 93150
(805) 969-4072
Dr. Richard Doutt

Retail

Plant Specialties: South African Bulbs and Seeds, especially summer-dormant bulbs from Cape Province, some hybridizing of Gladiolus, Amaryllis, Nerine, Brunsvigia; some pure species - Ixia, Freesia, Babiana, Lachenalia. Also evergreen species - Watsonia, Clivia, Dietes.

History/description: Another lucky victim of a rampaging hobby, Dr. Doutt discovered bulbs by way of a 30 year career as a UC Berkeley entomologist and environmental lawyer. In the bulb business since 1980, he has writen a book on Cape Bulbs, published in 1993. Currently he has returned to entomology and is studying of the fauna of the Channel Islands.

How to get there: Call for directions.

General information: Open by appointment only A catalogue is available and they will ship bulbs and seeds.

County: Santa Barbara

CARMEL VALLEY BEGONIA GARDENS

9220 Carmel Valley Road
Carmel 93923
(408) 624-7231
Todd Pascoe, James Flagg, Mark Eaton

Retail and Wholesale

Plant Specialties: Tuberous begonias, Geraniums (fancy-leaf and zonal), Fuchsias, Clematis (20 varieties), Orchids, Cyclamen, Azaleas, Perennials (many imported from Europe), Drought-tolerant plants, California native plants and Australian plants.

History/description: Carmel Valley Begonia Gardens is a full service grower. The present partners bought the business in 1989. The original owners were noted begonia growers, hence the nursery's name. However, the new direction is toward a more inclusive stock of rare and fascinating plants. While you are there, plan to visit the Lester Rowntree Arboretum in Carmel.

How to get there: From Highway 1 in Carmel, go six miles east on Carmel Valley Road.

General information: Open everyday, 8:30 am - 5 pm. They deliver to the Monterey Peninsula and will give slide talks to interested groups.

County: Monterey

CENTRAL COAST GROWERS

275 Pacific Pine Drive
Arroyo Grande 93420
(805) 489-1802
Dorothy and Roger Bunch

Wholesale

Plant Specialties: 800 varieties of herbaceous Perennials and Ground covers in 5-1/2" pots, including Salvias (80 varieties), Artemisias (13), Campanulas (27), Penstemons (25), Old-fashioned Roses.

History/description: Self-described as "a little guy with a huge inventory", Roger Bunch taught horticulture and agricultural mechanics while his children started a nursery to earn money during high school. In 1982 when the children wanted to move on to new ventures, the Bunches bought them out. They now have a mother stock of 4,000 5-gallon plants. They rotate their selections each year, testing new plants for 1-2 years before their official introduction. They recommend a visit to Cal Poly's Arboretum while in the area.

How to get there: Call for directions.

General information: Open to wholesale trade only, Monday through Friday, 8 am - 5 pm. Retail customers may visit with landscape professional. They deliver locally or can send a whole truckload almost anywhere. Catalogue is available upon request

County: San Luis Obispo

CHIA NURSERY

6380 Via Real
Carpinteria 93013
(805) 684-3382
Robert Abe

Retail and Wholesale

Plant Specialties: Ornamental Grasses, Bamboos, California native plants, Succulents, Perennials, Drought-tolerant plants, Herbs, big collections of Geranium, Cannas, Salvias.

History/description: Robert Abe was well-placed to buy this nursery of rare and unusual plants when former owner, Daryll Combs, split for greater spaciousness in Oregon. Robert's father owns Abe's Nursery next door. This 25 acre location includes a display garden, complete with carp pond and casitas de lizardos. He can give a garden talk to your group.

How to get there: Highway 101 south to Bailard Avenue exit in Carpinteria. Go left across freeway to Via Real. Go right on Via Real for 1/4 mile.

General information: Open Monday through Saturday, 8 am - 4:30 pm. Sunday, by appointment only. Plant list available for S.A.S.E.

County: Santa Barbara

DESERT THEATRE

17 Behler Road
Watsonville 95076
(408) 728-5513
Kate Jackson

Retail and Wholesale

Plant Specialties: Cacti and Succulents, with special collections of Euphorbia (145 varieties), Notocactus, Mammillaria, Haworthia, Echeveria, Gymnocalycium, Rebutia.

History/description: On vacation from her work as midwife - delivery nurse, Kate Jackson visited the southwest and fell in love with the desert. She began collecting a few plants. By 1982 she was selling her growing collection part-time and by 1987 she had a full-scale business. Worth a visit for the artistic arrangement of display beds, the nursery really is a re-created desert 'theatre'. She will pass along her love of plants by giving talks or tours of the nursery to groups.

How to get there: From Highway 101, take Highway 17 west to Highway 1 south. Drive 14 miles to Watsonville. Take Airport Boulevard exit to Green Valley Road. Go left Entrance to nursery is 1/2 mile on right.

General information: Hours vary with the season. Call for exact times. Closed Monday. They will ship mail orders. Catalogue is available for $2.

County: Santa Cruz

DROUGHT RESISTANT WHOLESALE NURSERY

Cypress Lane & Carmel Valley Road, or P. O. Box 1471
Carmel Valley 93923
(408) 624-6226, FAX (408) 624-0507
Thom Crow, Steve Halvorson

Wholesale

Plant Specialties: Drought-tolerant plants, Mediterranean plants, (special collections of Lavandula, Cistus), Australian, South African, and California native plants, including 16 varieties of Ceanothus, 8 of Arctostaphylos. Also Trees in 15-gallon cans.

History/description: In 1987 Thom and Steve saw the need for a first-rate wholesale producer of drought-tolerant plants. People like this give people like us infinitely more planting possibilities.

How to get there: Call for directions

General information: Open to wholesale trade. Public may come only with a landscape professional. Availability list for wholesale buyers.

County: Monterey

DYNASTY GARDENS

3621 A Main Street
Soquel 95073
(408) 475-5296 (Allow many rings.)
Robert Herse; Carol Herse, Manager

Retail and Wholesale

Plant Specialties: 600 varieties of unusual small-flowering Perennials and Alpines, especially Campanula, Salvia, Lavandula, Dianthus, Phlox, Thyme. Also Succulents, Aquatic plants, Ornamental Grasses, Bulbs, Annuals.

History/description: During his years as an Elementary School principal, Robert Herse was part owner of a cactus and succulent nursery. In 1983, after he left teaching, he started a nursery specializing in ground covers; over time his interests switched to perennials. The nursery is now managed by his biologist-daughter who emphasizes the importance of good growing ground. A soil specialist, she can help you with your soil problems and will do contract growing.

How to get there: From Highway 1, take Soquel/Capitola exit and go underneath overpass to stop sign. Go right on Main Street for one mile.

General information: Open Monday through Friday, 8:30 am - 5 pm, Saturday, 10 am - 2 pm. Not open Saturday, November to March. A plant list is available for S.A.S.E.

County: Santa Cruz

FARWEST NURSERIES
2669 Mattison Lane
Santa Cruz 95062
(408) 476-8865
Harry Petrakis

Retail and Wholesale

Plant Specialties: California native plants, Trees, large specimens and smaller sizes (1 gallon), general landscape plants, some Tropicals.

History/description: Farwest opened in 1973 on the site of a former nursery. Harry Petrakis and his former partner brought years of horticultural expertise to this undertaking. A quarter acre of their two and a quarter acre area is a display garden of perennials and bedding plants.

How to get there: Take Highway 17 to Highway 1 south. Take Soquel Avenue (second exit) to Mattison. Turn right onto Mattison.

General information: Open weekdays, 8 am - 4:30 pm; Saturday, 9 am - 4:30 pm; Sunday, 10 am - 4:30 pm. They will deliver locally. Catalogue available for wholesale customers only.

County: Santa Cruz

THE GERANIUM HOUSE

1331 Tunnel Road
Santa Barbara 93015
(805) 682-4175
Harriet Foster

Retail

Plant Specialties: Pelargoniums only - Zonal (100 varieties), Ivy (50 varieties), Scented (50 varieties).

History/description: Another glorious example of a hobby gone rampant. In 1968 this executive secretary moved to Santa Barbara, took some hort classes and discovered she had a knack for growing things. She started with trees but bent to her husband's suggestion that she try something smaller. She will guide you through her quarter acre garden to show you her pelargoniums and few other "oddball" things.

How to get there: Call for directions.

General information: Open by appointment only.

County: Santa Barbara

HAVER'S RHODENDRONS

2400 Bean Creek Road
Scotts Valley 95066
(408) 438-4090, FAX (408) 438-6070
The Haver Family

Retail and Wholesale

Plant Specialties: Rhododendrons; standards and some dwarfs.

History/description: For many years, the Haver family lived nearby this rhododendron nursery and so loved the property that when it came up for sale, they bought it. None of the eight family members presently working in the nursery had any horticultural background, so plant care experts were hired and crash learning courses begun. That was in 1981. Mrs. Haver, her children and their spouses still love the area and the rhododendrons flourish under their care.

How to get there: Take Highway 101 south to Highway 17 south. Take Scotts Valley exit. Go right on Glenwood Drive for 2.5 miles. Go left on Bean Creek Road. From Santa Cruz, take Highway 1 to Highway 17 north to Mt. Herman exit. Right on Scotts Valley Drive. Left on Bean Creek.

General information: Open Monday through Saturday, 8 am to 5 pm. They deliver plants FOB with minimum order.

County: Santa Cruz

HI-MARK NURSERY, INC.
1635 Cravens Lane
Carpinteria 93013
(805) 684-4462
Mark Bartholomew

Wholesale

Plant Specialties: Herbaceous, hardy Perennials, unusual Annuals, Begonias, Orchids, South African Bulbs in 4" to gallon sizes.

History/description: We would like to think this nursery, founded by Mark's father, R. C. Bartholomew, was named for the words of joy uttered at his son's birth. Not so. Having started in 1951, the nursery pre-dates Mark. Originally specializing in annuals, Hi-Mark made the big shift to perennials in 1984. They have a demonstration garden and offer tours by special arrangement.

How to get there: Call for directions.

General information: Although this business is primarily wholesale, public may come with landscape professional, weekdays, 7:30 am - 4 pm. A plant list is available for S.A.S.E.

County: Santa Barbara

HOPE RHODODENDRON NURSERY

264 Sims Road
Santa Cruz 95060
(408) 439-8959
Bruce Hope

Wholesale

Plant Specialties: Rhododendrons, container grown, heavily budded, two gallon size; especially Maddenii and fragrant types. Hundreds of other standard varieties. Dahlias (40 varieties) and other flowering perennials.

History/description: Studies at Cal Poly, San Luis Obispo convinced nurseryman Bruce Hope to choose a crop well-suited to the Santa Cruz area. In business since 1983, Hope does contract growing.

How to get there: Call for directions.

General information: Open by appointment only. Plant list available for S.A.S.E.

County: Santa Cruz

JOHN EWING ORCHIDS

487 White Road, or P. O. Box 1318
Soquel 95073
(408) 684-1111
Loraine and John Ewing

Retail and Wholesale

Plant Specialties: Orchids, specializing in the hybridizing of Phalaenopsis. Their hybrids include P. 'Mahalo' and P. 'Fire Water'. Some Cattleyas and Cymbidiums.

History/description: John Ewing started growing orchids at age nine, tending his first plant, a gift from his father. With degrees in botany and biology, and joined by an equally interested wife, he founded the nursery in 1969, located first in southern California and in Soquel for the past 20 years. Well-known hybridizers, they seek to extend the color range of Phalaenopsis, to include shades of pastels, peach, and stripes. They give lectures and will contract grow for clients.

How to get there: From Highway 101, take the Larkin Valley exit. Go east for 1 mile. Turn left onto White Road.

General information: Open Wednesday through Saturday, 9 am - 4 pm. They ship plants around the world. Their catalogue is free.

County: Santa Cruz

LAS PILITAS NURSERY

Star Route 23X
Santa Margarita 93453
(805) 438-5992
Celeste and Bert Wilson

Retail and Wholesale

Plant Specialties: 500 species of California native plants, especially Arctostaphylos.

History/description: Bert Wilson's landscape contracting business led to a fascination with native plants and an awareness of their unavailability. So the Wilsons switched to growing plants in 1975, aided by degrees in biology and botany. They moved to this 36 acre site in 1977 because its slightly acidic soil of decomposed granite is perfect for growing California natives. Time now limits his landscape contracting to botanical gardens.

How to get there: Take Highway 101 to Santa Margarita. Go east on Highway 58. After about a mile 58 bcomes Pozo Road. Continue for 3.5 miles. Left on Las Pilitas. Nursery is in 6 miles. Do not give up. Las Pilitas means "hole in the rock filled with water" which describes the road, not the nursery.

General information: Open Saturday, 9 am - 5 pm, but call first. Wholesale only, weekdays, 8 am - 6 pm. They will ship plants, Price list is available for S.A.S.E.

County: San Luis Obispo

LOS OSOS VALLEY NURSERY

301 Los Osos Valley Road
Los Osos 93402
(805) 528-5300
Tish and Lee Linsley

Retail

Plant Specialties: California native plants, Cacti, Mediterranean and Australian plants, drought-resistant Fuchsias, Indoor plants, Trees and Shrubs.

History/description: Although Los Osos sells plants grown by other growers (particularly Native Sons Nursery and San Marcos Growers), it does grow half of its stock on-site, and so we include it in this directory. There is no doubt that what it does sell is of special interest to plant lovers. The nursery began on an impulse. The Linsley's home adjoined a recently closed nursery. Rather than let the nursery license expire under a six month abandonment clause, they quickly bought it and have been expanding ever since. The 5-acre nursery has a display garden and Los Osos can help you plan yours.

How to get there: From Highway 101, take Los Osos exit going west. This is Los Osos Valley Road. Go 13 miles until the old farm on the corner of Los Osos Valley Road and Pecho Valley Road.

General information: Open everyday, 9 am - 5 pm. Closed Christmas and New Year's Day.

County: San Luis Obispo

LYON TREE FARM

2 Scarlett Road
Carmel Valley 93524
(408) 659-3196
Gary Townsend

Retail and Wholesale

Plant Specialties: Trees - Conifers (seedlings), 14 varieties of Eucalyptus. California native shrubs.

History/description: Gary Townsend started this primarily wholesale nursery in 1975, as a specialist in bare-root trees. Now he sells mostly plants as seedlings in supercells, focusing on volume sales to large buyers, such as reforestation projects and Christmas tree farms. He has some plants in 1 and 5 gallon sizes and will do contract growing.

How to get there: From Highway 1, take Carmel Valley Road for 6.5 miles. Right on Scarlett Road (half mile past Mid-Valley Shopping Center).

General information: Open Monday through Friday, 8 am - 5 pm. They will ship plants. Plant lists are available.

County: Monterey

MEADOWLARK NURSERY

824 Las Viboras Road
Hollister 95023
(408) 636-5912
Claire Steede-Butler

Retail and Wholesale

Plant Specialties: Salvias and unusual Drought-tolerant plants and California Native Perennials, Hardy Perennials.

History/description: As a landscape contactor, Claire Steede-Butler was frustrated by the short supply of perennials for her drought-resistant gardens. In 1986 she started this nursery in a beautiful valley at the southern extension of the Diablo Range because of its relatively cooler climate than that of the surrounding areas. Successfully growing perennials in this picturesque location, she soon hopes to have species and shrub roses for sale. She will give slide talks for small groups, can custom grow for special projects, and continues to build gardens in San Mateo and Santa Clara counties.

How to get there: From intersection of Highway 152 (Pacheco Pass Highway) and Highway 156, go south on Highway 156. Turn left on Fairview Road. Take second road on left which is Los Viboras Road. (called Churchill on the right). After 1 mile (approx.) nursery on unnamed side road to left.

General information: Open by appointment only. Deliveries to San Francisco, Peninsula, East Bay, Santa Clara, Monterey and Santa Cruz counties. Minimum order $250. Free 4-page plant list of hardy perennials is available.

County: San Benito

NATIVE SONS WHOLESALE NURSERY

379 West El Campo Road
Arroyo Grande 93420
(805) 481-5996
Dave Fross, Bob Keefe

Wholesale

Plant Specialties: California native plants, Drought-tolerant plants, Perennials, Native bunch grasses.

History/description: With a M.S. in agriculture, Dave Fross and partner Bob Keefe decided to open a nursery in 1979. Inspiration for its name came from their favorite Loggins and Messina song, "Native Sons". Since Fross and Keefe are both fourth generation Californians, they were bound to have an inborn fascination with California native plants. They suggest you also plan to visit the Cal Poly Arboretum, Nipomo Dunes and the Montana de Oro State Beach.

How to get there: Highway 101 South to El Campo Road exit. Go 1.5 miles to stop sign. Left on Los Berros. After 10 yards, go right on El Campo. Nursery in .9 miles.

General information: Wholesale customers only. Monday through Friday, 7:30 am - 4:30 pm; Saturday, 8 am - noon. Retail customers may visit only with a landscape professional. They deliver plants with a minimum order, produce a catalogue, and will give talks to groups.

County: San Luis Obispo

THE ORCHID HOUSE

1699 Sage Avenue
Los Osos 93402
(800) 235-4139
N. H. Powell

Retail and Wholesale

Plant Specialties: Orchids - Paphiopedilum, Phalaenopsis, Odontoglossum, many hybrids and cultivars of Paphiopedilum.

History/description: Trading bad habits for good, N. H. Powell started growing orchids when he gave up smoking fifty years ago. A Cal Tech-trained engineer, he found orchid raising to be a superb stress management technique. His scientific background has well served the orchid world and his efforts have received many awards. He claims the orchid of the future will be petite, easy to care for, long-lasting and a sure bet to flower every year. We are waiting. Largely wholesale, he welcomes seriously interested retail customers only.

How to get there: Call for directions.

General information: Open by appointment only, Monday through Saturday, 8 am - 5 pm. They will ship plants. Catalogue with directions for orchid culture available for $5.

County: San Luis Obispo

ORCHIDS OF LOS OSOS

1614 Sage Avenue
Los Osos 93402
(805) 528-0181
Michael Glikbarg

Retail and Wholesale

Plant Specialties: Orchids for indoor and outdoor use, especially cool-growing types. Cymbidium, Paphiopedilum, Pleione, Masdevallia, Zygopetalum and species orchids.

History/description: After graduation in ornamental horticulture from Cal Poly at San Luis Obispo, Michael Glikbarg bought out George Moran's orchid and cut flower business. Operating on these premises since 1964, he maintains one acre of greenhouses, offers site-specific consultations and will custom grow plants for customers. The native plant study area in Montana de Oro State Beach and the Cuesta College Botanical Gardens are of particular local interest for plant lovers. He alerts us to the soon-to-open botanical garden in Chorro Park. For more information write to Friends of SLO Botanical Garden, PO Box 4957, San Luis Obispo 93403.

How to get there: From Highway 101 just south of San Luis Obispo, take Los Osos-Baywood Park exit. Follow Los Osos Valley Road for about 10 miles. Go right on South Bay Boulevard. Go right on Nipomo. Go left on Sage to end.

General information: Open Tuesday through Friday, 9 am - 4 pm; Saturday, 11 am - 3:30 pm. They will ship and deliver plants, offer tours of the nursery and put on an orchid show in November. Catalogue is available for $3 which includes a $5 coupon that can be applied to any future purchase.

County: San Luis Obispo

PIANTA BELLA

3798 Via Real
Carpinteria 93013
(805) 969-7050
Bernard Acquistapace

Retail and Wholesale

Plant Specialties: Perennials, Ferns, Succulents, especially Aloe, Agave, Echeveria.

History/description: Thinking he might just earn a little extra money by selling plants to supplement his high school English teacher's salary, Bernard Acquistapace suddenly found himself in the nursery business. In 1986 he expanded his wholesale operation to include retail business. Needing still more space, he moved to this new location in 1993.

How to get there: Call for directions.

General information: Open Monday through Saturday, 8 am - 4:30 pm. Sunday, by appointment only. They will ship plants.

County: Santa Barbara

THE ROSE RANCH

240 Cooper Road, Salinas 93908
P. O. Box 10087, Salinas 93912
(408) 758-6965
Alice Hinton, Anthony Hinton

Retail

Plant Specialties: Over 800 varieties of Old Garden Roses, especially Centifolia, Damask, Bourbon. Some Hybrid Tea and Floribunda.

History/description: Alice Hinton's father grew old roses for 65 years and she shared the family passion. When he became ill, she took over. Joined by her son, she opened for business in 1992. Situated in the middle of lettuce fields, they offer a Spring open house and will give talks and demonstrations. The father's lovingly tended roses are their display garden. They suggest you also visit the Steinbeck House in Salinas and Garland Ranch in Carmel Valley.

How to get there: From Highway 101 south, take the Monterey Road exit. Go east over the overpass onto Molera Road. Molera Road becomes Cooper Road.

General information: Open Monday through Friday, 9 am - 5 pm. Sunday and Saturday, call first. They will ship plants. Plant list available for S.A.S.E.

County: Monterey

ROSENDALE NURSERY

2660 E. Lake Avenue
Watsonville 95076
(408) 728-2599
Lisa and Jeff Rosendale

Wholesale

Plant Specialties: Australian, South African and California native plants, Ferns (extensive selection), wide variety of Mediterranean and ornamental plants, special collection of Protea and Grevillea.

History/description: In 1989 the Rosendales had the opportunity to buy an existing nursery business after having been involved professionally with plants for 15 years. Their six acre site, located at the base of the coastal foothills east of Pajaro Valley near Hecker Pass, has a temperate, coastal climate conducive to growing a great variety of plant material. A demonstration garden is being developed.

How to get there: East from Highway 1 or west from Highway 101 to Highway 152 (Hecker Pass). Nursery located 2 miles east of Watsonville on Highway 152 across from the Santa Cruz County Fairgrounds.

General information: Open to wholesale trade only. Visit with your landscaper. They deliver throughout the San Francisco and Monterey Bay Areas.

County: Santa Cruz

ROSES OF YESTERDAY & TODAY

803 Brown's Valley Road
Watsonville 95076
(408) 724-3537, (408) 724-2755,
FAX (408) 724-1408
The Wiley Family

Retail

Plant Specialties: Old, rare and unusual Roses.

History/description: This is the grandddaddy of the rose business in California, having been around since 1935. The display garden has about 400 old, rare and unusual roses in full bloom in May and June. The Wiley family's love of roses is everywhere evident.

How to get there: Watsonville is 2 hours south of San Francisco, 30 miles north of Monterey and 17 miles south of Santa Cruz. Nearest major highway is Highway 1. Call for more complete directions.

General information: Display Garden is open 9 am - 3 pm, everyday. Office is open 9 am - 3 pm, Monday - Friday. Small gift shop in office. They ship worldwide. Handsome catalogue available for $3. They give slide talks for interested groups of 35 or more.

County: Santa Cruz

SAN MARCOS GROWERS WHOLESALE NURSERY

125 South San Marcos Road
P. O. Box 6827
Santa Barbara 93111
(805) 683-1561, FAX (805) 964-1329
Randy Baldwin

Wholesale

Plant Specialties: Drought-tolerant plants, Perennials, California native plants, Mediterranean plants, Australian plants, South African plants.

History/description: Since 1980 San Marcos has been actively introducing water-conserving plants from mediterranean areas around the world. They seek out plants which have horticultural merit, ornamental value, and are appropriate to the micro-climates of California. They have exclusively propagated the new color hybrids of New Zealand flax and can take a good deal of credit for the lush look now possible in drought-tolerant gardens. Their 21 acre location includes demonstration gardens and a full acre of propagation greenhouses.

How to get there: Call for directions.

General information: Open to wholesale customers only. Monday - Friday, 7:30 am - 4:30 pm; Saturday, 8 am - noon. They will ship and deliver. Catalogue available for the wholesale nursery trade.

County: Santa Barbara

SAN SIMEON NURSERY

Cayucos 93430
(805) 995-2466
John Goetz

Wholesale

Plant Specialties: Low water use plants. Australian plants, including Melaleuca, Coffea, Luma, Azara dentata, Casuarina stricta. Mediterranean plants, California native plants, Perennials, Palms, Cacti and Succulents, including a large variety of Aloes, Dudleyas, and Agaves (up to 24" box).

History/description: This owner-contractor specializes in low water-use landscapes. He opened the nursery in 1980 to make more plants available for his landscapes and yours. With degrees in horticulture and botany and years of practice in the field, he is well placed to enrich the horticulture of the Central Coast. He offers site-specific consultations and will give talks about his specialties.

How to get there: Call for directions.

General information: Open to wholesale customers only, by appointment. Public may come with a landscaper.

County: San Luis Obispo

SANTA BARBARA ORCHID ESTATE

1250 Orchid Drive
Santa Barbara 93111
(805) 967-1284
Anne Gripp

Retail and Wholesale

Plant Specialties: Orchids - Cattleya, Cymbidium, Phalaenopsis, Lycaste, Dendrobium, largest collection of species orchids, some hybrids.

History/description: Santa Barbara Orchid Estate, in business since 1957, was bought in 1967 by Anne Gripp from the original owner for whom she worked. A trained horticulturist, she has been expanding and refining the business ever since. Although the nursery has introduced many named hybrids, such as Lycaste 'Paul Gripp' and L. 'Angulo-caste', it now concentrates on self-propagating and growing species plants to preserve endangered species. The many orchid houses are open to visitors, and they offer on-site potting and orchid care workshops. She suggests you combine your visit with a trip to the Santa Barbara Botanical Garden, Alice Keck Memorial Park, the plantings of the Santa Barbara Zoo, and the Rose Garden across from the mission.

How to get there: Just north of Santa Barbara on Highway 101, take Patterson exit and head towards the ocean. After an agricultural area and at a bridge, the road goes left. Follow the yellow line, Patterson has now become Shoreline. Go right on Orchard Drive past the No Trespassing sign. Go left after third speed bump.

General information: Open Monday through Saturday, 8 am - 5:30 pm; Sunday, 10 am - 4 pm. They ship plants. Plant list is free.

County: Santa Barbara

SANTA BARBARA WATER GARDENS

160 East Mountain Drive
Santa Barbara 93140
(805) 969-5129
Virginia Hayes, Stephne Sheatsley

Retail

Plant Specialties: Bog plants, Aquatic plants, including Lotus, Nymphaea, hardy and tropical. Equipment necessary to install a water garden.

History/description: Stephne Sheatsley started the nursery in 1978 with an old horse trough and one piece of water lily (Chromatella) from her aunt's pond. Virginia Hayes joined her growing concern in 1986. Their collection has quadrupled in the last ten years, propagated primarily by vegetative division and some from seed. They have two demonstration ponds on site, will contract grow, and can help you design your water garden.

How to get there: From Highway 101, take the Milpas exit. Head toward the mountains and go right on Montecito Street. At roundabout, take Sycamore Canyon Road for 1.5 miles. At stop sign, take Coyote Road up hill to dead end. Go right on Mountain Drive.

General information: Open Wednesday, 8:30 am - 4:30 pm; Saturday, 10 am - 4 pm or by appointment. They ship plants. Catalogue available for $2, refundable with purchase.

County: Santa Barbara

SHEIN'S CACTUS

3360 Drew Street
Marina 93933
(408) 384-7765
Anne and Rubin Shein

Retail

Plant Specialties: Cacti and Succulents, especially Mammillaria, Copiapoa, Neochilenia, Neoporteria, Rebutia. Also Haworthia specialties and other rare and unusual plants.

History/description: Rubin Shein began collecting cacti and succulents as a child growing up in Germany. Arriving in the US in 1951, his excitement at seeing these plants growing in their natural habitat caused him to collect in earnest. Twenty six years of plant collecting produced quite a start-up inventory for this business which officially opened in 1977. Now there are seven greenhouses and an outdoor display garden of cold-tolerant cactus.

How to get there: Call for directions.

General information: Open by appointment only. They will ship plants and give talks. Plant list is available for $1.

County: Monterey

STEWART ORCHIDS

3376 Foothill Road
Carpinteria 93013
(805) 684-5448, FAX (805) 566-6609
William Stewart; Ned Nash, President

Retail and Wholesale

Plant Specialties: Orchids - Cattleya, Cymbidium, Phalaenopsis, Paphiopedilum, Miltonia.

History/description: The Stewart family has been in the orchid business since 1908. The two acres of greenhouses in Carpinteria have been part of the picture since 1977. The present company was formed by the merger of three orchid growers in 1985. Stewart Orchids is the only orchid nursery to use the patent process. They have 15 patented varieties, including Laeliacattleya 'Puppy Love' and L. 'Prism Palette'. Tours can be booked by prior arrangement.

How to get there: From Highway 101 south, take Padaro Lane exit. Go back over freeway, then right (south) on frontage road. Go left at Polo Field on Nidever. Nidever runs into Foothill.

General information: Open Monday through Friday, 8 am - 4 pm. Saturday, 10 am - 4 pm. They will ship plants. Free catalogue is available.

County: Santa Barbara

SUNCREST NURSERY

400 Casserly Road
Watsonville 95076
(408) 728-2595
Stan Iversen: Jim Marshall, Manager

Wholesale

Plant Specialties: All five Mediterranean climate zone plants, California native plants especially large Arctostaphylos collection, Perennials, Shrubs, and Trees.

History/description: Although Suncrest started in 1991, it is a synthesis of revered California horticultural enterprises. On the 40 acre site of the former Leonard Coates nursery, in business since 1876, it combines the expertise of this institution with the enviable stock (largely Australian) of Wintergreen Nursery. Newly constituted, it remains one of the biggest concerns in this specialty area.

How to get there: From Highway 101, go west on Highway 152 to bottom of hill. Go left (south) on Casserly Road.

General information: Open to wholesale trade ony, Monday through Friday, 8 am - 4:30 pm. Retail customers may visit only with a landscape professional. They deliver locally. Plant list is available upon request.

County: Santa Cruz

SUNSET COAST NURSERY

P. O Box 221
Watsonville 95077
(408) 726-1672
Patty Kreiberg

Retail and Wholesale

Plant Specialties: California native plants suitable to the coast, especially to the dunes and beaches of Monterey County. Salt marsh revegetation plants. Their collections are growing to include species suited to bluff, prairie, and oak woodland areas.

History/description: Trained as a biologist, Patty Kreiberg was a home gardener with an enviable crop of winter flowers and vegetables. Her hobby got serious when she learned from a friend about the lack of native plant material available for revegetation projects underway to repair the coastline damaged by winter storms in 1983. Her first sale was to the Monterey Bay Aquarium. Mother Nature's new best friend, she does contract growing and has a first rate slide show available for interested groups. Plan to visit nearby Mt. Madonna Park and Elkhorn Slough National Estuarine Reserve.

How to get there: Call for directions.

General information: Open anytime, by appointment only. Plant list available for S.A.S.E.

County: Santa Cruz

TAYLOR'S GREENWOOD NURSERY

2 El Camino Ratel
Goleta 93117
(805) 964-2420
Jane and Ken Taylor

Retail and Wholesale

Plant Specialties: 200 species and cultivated varieties of Cacti and Succulents, all cold-hardy, including Euphorbia obesa, Astrophytum myriostigma, Mammillaria, Opuntia, Parodia, Gymnocalycium. Palms, including Kentia, Livistona decipiens, L. mariae, L australis, Chamaedorea, Ravenala rivularis, Rhopalostylis baueri, R. sapida.

History/description: A fifth-generation southern Californian, Ken had a geographic affinity for palms. Jane became intrigued when she moved west. Together they started to collect seed of hardy palms and their collections grew. They still marvel at the myriad forms and geometries of their cacti, succulents and palms, and at the mechanisms they have developed in the wild for their survival. They consider it their challenge to recreate natural habitat conditions in container culture. They offer plant consultations at your site and have a display garden.

How to get there: Call for directions.

General information: Open by appointment only. Plant list is available for S.A.S.E.

County: Santa Barbara

■200 Where On Earth!

TIEDEMANN NURSERY

4835 Cherryvale Avenue
P. O. Box 926
Soquel 95073
(408) 475-5163, FAX (408) 475-4067
Jon Beard, General Manager

Wholesale

Plant Specialties: California native plants (100 varieties, including large Ceanothus collection), Perennials, Fuchsias, Cyclamen, Poinsettias.

History/description: This nursery has been growing plants for 40 years, under the tutelage of its present family since 1975. Nearby horticultural attractions include UCSC Arboretum and Butano State Park.

How to get there: Call for directions.

General information: Open to wholesale customers, Monday to Friday, 7:30 am - 4 pm. Public visits only with landscape professional. They will deliver from Santa Rosa to Carmel and will give slide talks to groups.

County: Santa Cruz

WILLIAM R. P. WELCH

43 E. Garzas Road (mailing address),
264 West Carmel Valley Road (growing fields)
Carmel Valley 93924
(408) 659-3830 -office
(408) 659-0588-field telephone, allow many rings
William R. P. Welch

Retail and Wholesale

Plant Specialties: Cluster-flowered Narcissus, Daffodils and Amaryllis (Belladonna hybrids).

History/description: As a young lepidopterist, William Welch began to grow plants as food for his caterpillar collections. Still in his late teens he got into the nursery business in 1979, selecting narcissus as his specialty because they were well suited to his area, deer and gopher resistant, and reasonably drought tolerant. Located in the upper bench area of the Carmel Valley, he welcomes visitors to his 5 acre growing fields to enliven his somewhat solitary horticultural pursuits. During flowering season, he will gladly sell cut flowers and will give talks anytime.

How to get there: From Highway 101, take Highway 68 toward Monterey. Go left on Los Laureles Grade Road. Left on Carmel Valley Road for one mile. Nursery on right hand side of road opposite Country Club Drive.

General information: Open by appointment only. Flower season: Narcissus and Daffodils - November to April; Amarylis - August to September. Bulb Season (field digging): June through October. Bulbs may be ordered from mailing address. Free catalogue available.

County: Monterey

HORTICULTURAL ATTRACTIONS

Santa Barbara Botanic Garden

1212 Mission Canyon Road
Santa Barbara 93105
(805) 682-4726

History/description: The 65-acre grounds of the Santa Barbara Botanic Garden is dedicated to the display, propagation, and preservation of California's native flora. Featured areas include a Demonstration Garden showing landscape use of native plants, and Meadow, Redwood Forest and Channel Island sections. Docent tours of the garden are held everyday at 2 pm and Thursdays and weekends at 10:30 am. Special focus tours given every other Monday at 12:10. Wildflower celebration held in the spring.

General information: Garden open everyday, 8 am to sunset. Library and office open weekdays, 9 am - 4:45 pm (closed for lunch). Nursery open 9 am - 5 pm (9-4 in winter).

Plant sales: The volunteer Garden Growers manage the nursery to raise income for the Garden. Plants are sold through their shop and in big Spring and Fall plants sales.

UC Santa Cruz Arboretum

University of California
Santa Cruz 95064
(408) 427-2998

Central Coast

History/description: UC Santa Cruz Arboretum specializes in Southern Hemisphere drought tolerant plants and has the largest collection of Australian, New Zealand and South African proteas outside each of these countries. Fifty of their 200 acres are planted. Arboretum Associates, the membership support group, organizes lectures, and slide shows. Group tours may be arranged.

General information: Open from 9-5 every day.

Plant sales: Plants are for sale, Wednesday, Saturday and Sunday, from 2-4 pm. Major plant sales occur on the third Saturday in May and the second Saturday in October.

Other Attractions:

Leaning Pine Arboretum

California Polytechnic State University
San Luis Obispo 93407
(805) 756-2279

Two acres, part of Ornamental Horticulture Unit which includes greenhouses, labs and nursery selling student-propagated plants.

Lotusland

695 Ashley Road
Montecito 93108
(805) 969-3767

Scheduled to open Fall, 1993; reservations required. Opera-diva Ganna Walska's estate with cactus, cycad and succulent gardens.

Lester Rowntree Arboretum

25800 Hatton Road
Carmel
(408) 624-3543

Native garden honoring California naturalist; part of Mission Trail Park.

■204 Where On Earth!

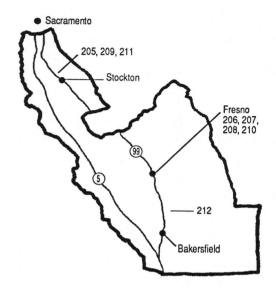

SAN JOAQUIN VALLEY

**SAN JOAQUIN
FRESNO
TULARE**

CACTUS BY DODIE

934 East Mettler Road
Lodi 95242
(209) 368-3692
Dodie and Dick Suess

Retail

Plant Specialties: Cacti and Succulents, rare stock from seed including Haworthia, Crassula, Conophytum, Euphorbia, and large cactus collection.

History/description: Dodie Suess was a cactus and succulent hobbyist, as were her parents. She and her husband started a retail nursery by buying the stock and mail order list of a going-out-of-business nursery in 1981. They have a small display garden, though most plants are in greenhouses.

How to get there: Call for directions.

General information: Open Thursday through Saturday, 9 am - 5 pm. They will ship orders. Mail order catalogue available for $2.

County: San Joaquin

HENDERSON'S EXPERIMENTAL GARDENS

1215 N. McCall
P. O. Box 612
Clovis 93612
(209) 251-8393
Don Kleim

Retail and Wholesale

Plant Specialties: Japanese Maples (150 varieties), deciduous Magnolias (109 varieties), Nandinas (12 varieties, including the true form of N. purpurea 'nana'), Dwarf Conifers, Kurume Azaleas, Camellias (40 Sasanqua varieties, 20 hybrids), Helleborus 'atrorubens', Ornamental Grasses.

History/description: Mr. Henderson, the last associate of Luther Burbank, started this nursery in 1926. Don Kleim worked with Mr. Henderson from 1946 until he took over the business in 1972. Extensive travel in Japan in the 1960s sparked an interest in Japanese maples. This 13 acre nursery on rich bottomland soil boasts the largest collection of deciduous magnolias on the west coast. He was the first to import Nandina 'Chiriman' and introduced N. 'Abundance' and Clematis armandii 'Hendersonii rubra'.

How to get there: Highway 99 south to Herndon Exit. Go east for 13 miles. Right (south) on McCall.

General information: Open Monday through Saturday, 8 am - 4:30 pm; Sunday 10 am - 4 pm. February through November. They deliver throughout the west and have a wholesale plant list available. No mail orders. They give talks to interested groups.

County: Fresno

HILL 'N DALE
6427 North Fruit Avenue
Fresno 93711
(209) 439-8249
Dale Kloppenburg

Retail and Wholesale

Plant Specialties: Primarily Hoyas, about 300 varieties.

History/description: Dale Kloppenburg has done more than most. With a degree in genetics from UC Berkeley, he started his career as a rose and fruit tree breeder. From 1957 to 1968 he opened and ran an arts and crafts store. Then came a stint as research agronomist for Northup King Seeds. In 1979 he followed the lure of his new hobby, hoyas, travelling to Australia, Singapore, Malaysia and elsewhere in search of all varieties of this genus. He has written *Hoyas of the Philippines* and *Hoya Handbook*, as guides to this versatile plant which is sometimes succulent, thin-leafed, vining, or even epiphytic.

How to get there: Call for directions.

General information: Open by appointment only. They will ship mail orders. Catalogue available for $1.

County: Fresno

KELLY'S PLANT WORLD

10266 E. Princeton Avenue
Sanger 93657
(209) 292-3505
Herbert Kelly, Jr.

Retail and Wholesale

Plant Specialties: Perennials, Canna, Crinum, Gardenias, Hymenocallis, Hedychium (Gingers), Siberian iris, Lycoris, plants and Bulbs from around the globe, cold-hardy Palms.

History/description: Hybridizer Herb Kelly started this nursery in the early 1980s to share his love of rare and unusual plants. He has created the first yellow hybrid crinum (Crinum 'Yellow Triumph'). He grows an incredible 125 varieties of canna, over 150 types of lycoris, 150 clones of Amaryllis belladonna, and more than 25 varieties of gardenias. All plants are field grown on six acres of land. More than 1,000 varieties of unusual plants make this family nursery worth a visit.

How to get there: Highway 99 south to Herndon Exit. Go left over railroad tracks to Herndon Avenue. Continue east for 13 miles through Fresno and Clovis to McCall Avenue. Right on McCall. After two stop signs go about one mile and turn left on Princeton.

General information: Open everyday, 8 am - 5 pm. They will ship plants and have a catalogue available for $1 (refundable on first order). Slide talks available for groups.

County: Fresno

KING'S MUMS

20303 E. Liberty Road, or P. O. Box 368
Clements 95227
(209) 759-3571
Ted King, Lanna King

Retail

Plant Specialties: Chrysanthemums - 250 different cultivars, the most complete selection in the U.S.A. Sold in 6-1/2" pots or as rooted cuttings in spring. Football types, spiders, cascades, bonsai types and cushions.

History/description: King's was started in 1964 in Castro Valley as a sideline to Ted King's pest control business. Hoping to locate fellow chrysanthemum hobbyists by starting the business, he was obviously successful. King's has become the largest producer of chrysanthemum cultivars. Their annual fall King's Mums Show (October 1 - November 24) is well worth a visit to see the fields of flowers growing on California's rolling Sierra foothills. A display garden illustrates the many styles of chrysanthemum culture - cascade, tree, bonsai, hanging basket and espalier. All is not lost if you pass by there in the spring; he suggests you visit Daffodil Hill near Volcano further east on Route 88.

How to get there: One mile west of intersection of Highway 88 and Liberty Road. 11 miles east of intersection of Highway 99 and Liberty Road.

General information: Open October 1 to November 24. They ship to U.S. and Canada. Color catalogue available for $2, deductible from purchase. Special lectures and workshops can be arranged.

County: San Joaquin

PLANET EARTH GROWERS NURSERY

14178 West Kearney Boulevard
Kerman 93630
(209) 846-7881
Cathy and John Etheridge

Retail and Wholesale

Plant Specialties: Australian plants, California native plants, South African plants, with large collections of Cistus, Penstemon, Mimulus/Diplacus, Grevillea, Acacia, and Eucalyptus

History/description: The Etheridges began experimenting with xeriscape plants when they opened their nursery in Arizona in the 1970s. Moving to California, they founded Planet Earth in 1987. On their four acres of growing fields, they maintain a one acre demonstration garden to test plants for their temperature tolerance (17 - 104 degrees).

How to get there: From Fresno take Highway 99 south to Shaw exit. Go west approximately ten miles. Left on Dickenson Road. Right on Kearney Blvd. Nursery is in four miles.

General information: Open Tuesday through Sunday, 9 am - 5 pm. Tours for groups may be arranged.

County: Fresno

SEIJU-EN BONSAI NURSERY

23181 North Davis Road
Lodi 95242
(209) 369-4168
Shin and Carl Young

Retail and Wholesale

Plant Specialties: Bonsai. Plants grown from seeds, cuttings, grafting, division, layering or from collected specimens. Field and container grown to develop caliper, the stock includes pre-bonsai, partially trained and finished bonsai.

History/description: Carl Young spent twenty years in Japan where he met and married Shin. Together they studied with the world's foremost bonsai masters, becoming registered teachers of bonkei (tray landscapes) and co-authoring a book on chrysanthemum bonsai. They introduced Ulmus parvifolia 'seiju', propagated from a sport. Plants range from seedlings to 17-gallon trained specimens. The nursery started in 1969 on a ten acre, former vineyard.

How to get there: Call for directions

General information: Open everyday, 8 am - 4:30 pm, but call for appointment. They will ship plants. Minimum order is $25. Catalogue is available.

County: San Joaquin

SEQUOIA NURSERY

2519 East Noble Avenue
Visalia 93292
(209) 732-0190
Ralph Moore

Retail and Wholesale

Plant Specialties: Roses, specializing in miniatures (owner-hybridized); also antique roses, species roses, old teas.

History/description: As long as he can remember, Ralph Moore has always been interested in roses. Son of a farmer, he established a general nursery of landscape plants in 1937 and planted the trees for which the nursery is named. By the 1950s the nursery was devoted entirely to roses. He has developed over 400 varieties of roses, mostly miniatures, registered by the American Rose Society. Most are very unusual, very floriferous, and often everblooming. He is still going strong.

How to get there: From Highway 99, take Highway 198 east for about 8 miles. Take Ben Maddox exit to Noble Avenue. Go left on Noble for 1/2 mile. You will see the Sequoias.

General information: Open Monday through Friday, 8:30 am - 4 pm; Saturday and Sunday, 10 am - 3 pm. They will ship plants. Mail order catalogue available for $1.

County: Tulare

HORTICULTURAL ATTRACTIONS

Micke Grove Park
Stockton (between Highway 99 and I-5)
(209) 463-0578

Oak grove, Japanese garden, camellia garden, rose garden.

Roeding Park
890 West Belmont
Fresno
(209) 498-1551
(510) 562-0328

Impressive, old, labelled trees.

■214 Where On Earth!

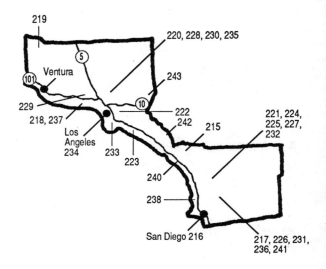

SOUTHERN CALIFORNIA

**RIVERSIDE
SAN DIEGO
LOS ANGELES
ORANGE
SAN BERNARDINO
VENTURA**

BLOOMING FIELDS FARM

P.O. Box 8416
Riverside 92515
(909) 352-1482
Jim Puckett

Retail and Wholesale

Plant Specialties: Tall-bearded Iris - 2,000 varieties, including rebloomers 'Champagne Elegance', 'Eternal Bliss'. Heritage iris (200 varieties), some from the 1840s.

History/description: Geologist Jim Puckett has been a serious iris collector for 35 years. Since 1988 his hobby has become a business; now geological consulting takes a back seat. Having an impressive collection of heritage iris, he can supply owners of period homes with irises to match. He lectures locally and is working on a display garden with Greenwood Daylily Gardens, whose growing grounds are next door.

How to get there: Call for directions.

General information: Open during April (bloom season), Sundays, 9 am - 4 pm. They ship plants. Mail order catalogue is free.

County: Riverside

BLOSSOM VALLEY GARDENS

15011 Oak Creek Road
El Cajon 92021-2328
(619) 443-7711
Patricia and Sanford Roberts

Retail

Plant Specialties: Daylilies, tetraploids only. Epiphyllum.

History/description: They could not take the farm out of this boy. Sanford Roberts grew his first daylily on his grandmother's farm in 1941. Not even 25 years of military service could quench his love of growing things. Now also a dog breeder and garden/sports writer, he has been growing daylilies in earnest on his three acre mountaintop home east of El Cajon since 1982. His home garden includes a 1,500 square foot shade house for his epiphyllums.

How to get there: Call for directions.

General information: Open by appointment only. They ship plants. Mail order catalogue available for $1.

County: San Diego

BUENA CREEK GARDENS

418 Buena Creek Road
San Marcos 92069
(619) 744-2810
Bob Brooks; Steve Brigham, Manager

Retail and Wholesale

Plant Specialties: Daylilies, Iris, Perennials, Subtropicals, Drought tolerant plants, Alstroemerias (division-grown), Anigozanthus.

History/description: Since 1970 Bob Brooks has had a successful mail order business, specializing in connoisseur quality daylilies (Cordon Bleu Farms). In 1988, wanting to create a public garden and retail nursery, he hired Steve Brigham, a horticulturist trained at UCSC, Kartuz Greenhouses, and Quail Botanical Gardens. The result is four acres of display gardens done in the lively, curvilinear, tropical style of Brazilian Burle Marx surrounding a beautifully-sited, 1930s ranch house.

How to get there: From Highway 5 south at Oceanside, take Highway 78 to Twin Oaks Valley Road. Go left for three miles through valley. Left on Buena Creek Road for 1/4 mile. From Highway 15, take Deer Springs Road exit to Twin Oaks Valley Road, then right on Buena Creek Road.

General information: Open Wednesday through Saturday, 9 am - 4 pm. Cordon Bleu Farms has a catalogue available ($1) for mail orders. Write to P. O. Box 2033, San Marcos 92079.

County: San Diego

COASTAL ZONE NURSERY

31427 Pacific Coast Highway
Malibu 90265
(310) 457-3343
Richard Howlett, Bill Long

Retail and Wholesale

Plant Specialties: Plants primarily suited to Central Coast and Southern California climate. Over 170 varieties of Trees, Palms, Ferns, Shrubs and coastal California native plants. Collections include Acacia, Calodendrum, Castanospermum, Corynocarpus, Corynodendron, Cussonia, Erythrina, Markhamia, Melaleuca (M. diosmafolia, M. squamia, M. ircana, M. decussara), Spathodea, Stenocarpus, Vitex lucens.

History/description: Bill Long bought this property in 1978, formerly the site of La Herran Nursery in the 1940s. He was inspired to start a tree nursery by the many old specimen trees still on this 30 acre site. Almost all of their business is wholesale. Dealing in sizes from 5 gallon to 60" boxes, they have cranes available for large installations. They recommend combining your visit with a trip to the Getty Museum gardens, the garden at Malibu Lagoon State Beach and Will Rogers Historic Park in Pacific Palisades.

How to get there: The nursery is located 3/4 mile north of intersection of Trancas Canyon Road and Pacific Coast Highway on east side of road.

General information: Open Monday through Friday, 8 am - 4 pm. They have a plant list for wholesale customers only.

County: Los Angeles

DESERT IMAGES

11140 N. Ventura Avenue
Ojai 93023
(805) 649-4479
Richard Bogart

Retail

Plant Specialties: Cacti, Succulents, related landscape plants and potted, indoor plants.

History/description: Desert Images began as a part-time diversion from Richard Bogart's full-time occupation with engineering/survey work in 1972. No doubt encouraged by his botanist-trained wife, he developed an interest in succulents. A full-time business since 1982, the nursery includes a display garden.

How to get there: From Highway 101 take Highway 33 towards Ojai. Highway 33 becomes Ventura Avenue. Nursery is one mile on the right after the town of Oak View on the right.

General information: Open weekdays, 10 am - 4 pm; weekends, 9 am - 5 pm.

County: Ventura

DESERT TO JUNGLE NURSERY

3211 West Beverly Blvd.
Montebello 90640
(213) 722-3976
Gary Hammer, Owner; Del Pace, Manager

Retail and Wholesale

Plant Specialties: Perennials for mediterranean and subtropical climates, Drought-tolerant plants from Australia and South Africa.

History/description: Desert to Jungle and its wholesale counterpart, Glendale Paradise Nursery in Lakeview Terrace, are responsible for countless introductions of rare and unusual plants to California. Many plant exploration trips to Australia, South Africa and South and Central America have unearthed unique perennials suited to both mediterranean and subtropical climates.

How to get there: Call for directions.

General information: Open Wednesday - Sunday, 10 am - 4 pm.

County: Los Angeles

EXOTICA RARE FRUIT NURSERY AND SEED COMPANY

2508B East Vista Way
Vista 92083
(619) 724-9093
Jessica Leaf, Steven Spangler

Retail and Wholesale

Plant Specialties: Edible landscape plants, especially Tropical fruits. Subtropical and Mediterranean Shrubs, fruiting, flowering and fragrant plants, Bamboo, Palms. All plants organically and bio-dynamically grown. Figs (30 varieties), old Guavas, Mulberries, tropical Cherries, sweet Pomegranates. Asian medicinal plants - longani, lychee, jujube. Seeds and compost for wholesale only.

History/description: Steven Spangler supplemented his oceanographic studies in Japan picking pineapples. He discovered a unique Tahitian squash and advertised its seeds in Organic Gardening. English seed company Thompson and Morgan ordered a huge amount and the Exotica Seed Company was born. Jessica Leaf joined the partnership in 1984, establishing the nursery as a miniature botanical garden on 2.5 acres. They give lectures and design consultations and will contract grow.

How to get there: From Highway 5 about 3/4 hour north of San Diego, go east for 6.5 miles on Highway 78. Exit on Escondido Avenue. At the sixth stop light, go right on East Vista Way for 2.5 miles.

General information: Open everyday, 10 am - 4 pm. They ship mail orders. Catalogue available for S.A.S.E.

County: San Diego

GREENLEE NURSERY

301 East Franklin Avenue
Pomona 91766
(909) 629-9045, FAX (909) 620-9283
John Greenlee; Jackie Coburn

Retail and Wholesale

Plant Specialties: Ornamental Grasses and grass-like plants.

History/description: Probably more than anyone, John Greenlee is responsible for the increased interest in ornamental grasses in the west. He is an articulate spokesman for his product and is generous with his expertise. Starting as the western representative for Kurt Bluemel in 1984, he has since become his own best spokesman for the splendors of grass. His recent book, *The Encyclopedia of Ornamental Grasses* offers color photographs and comprehensive descriptions of over 300 ornamental grasses.

How to get there: Call for directions.

General information: Open by appointment only. Mail order catalogue is available for $5.

County: Los Angeles

GREENWOOD DAYLILIES

5595 East 7th Street, #490
Long Beach 90804
(310) 494-8944, FAX (310) 494-0486
Cynthia and John Schoustra

Retail and Wholesale

Plant Specialties: Daylilies, nearly 2,000 different cultivars, field and container grown.

History/description: UC Berkeley-educated landscape architect, John Schoustra was a landscape contractor for ten years. He started a soil business because he could not find good dirt for his clients. He started the nursery when named daylilies, well-suited to southern California, were not locally available. The Schoustras bought their initial stock of daylilies from the Greenwod Nursery in Goleta when it switched its specialty. Shirley Kerins of Huntington Botanical Gardens designed the color layout of their display garden. John is on the brand-new All-American Daylily Selection Council which will field test daylilies and, thus, get them into the general nursery trade. The Schoustras give tours of their growing grounds in Riverside, give talks and lectures, and can help landscape architects with their plant selections. They will contract grow.

How to get there: Take Highways 405, 605, or 22 to 7th Street/Long Beach exit. Take the Studebaker Road offramp to go left (south) on Studebaker Road. Go left at next light. Go a few hundred yards, then right into nursery.

General information: Open the first and third Saturday of each month except December and January or by appointment. They will ship plants. Illustrated poster and catalogue available for $5, refundable from purchase.

County: Los Angeles

GRIGSBY CACTUS GARDENS

2354 Bella Vista Drive
Vista 92084
(619) 727-1323
David Grigsby; Madelyn Lee, Manager

Retail and Wholesale

Plant Specialties: Cacti and Succulents, including Euphorbia, Sansevieria, large specimen plants for collectors and landscape use.

History/description: David Grigsby had been collecing cacti and succulents for years. In 1965 he bought his present land to retire on and started planting out his "hobby plants". The hobby ran amok and resulted in their very special 3 acre display garden and nursery. They give tours of the garden for groups. All plants are artificially propagated; none is wild collected.

How to get there: Call for directions.

General information: Open Tuesday through Saturday, 8 am - 3:30 pm. They will ship plants. Catalogue available for $2.

County: San Diego

HERBAN GARDEN HERBS

5002 2nd Street
Fallbrook 92028
(619) 723-2967
Jeanne and Chris Dunn

Retail and Wholesale

Plant Specialties: 100 varieties of Herbs, including Thyme (10 varieties), Mint (13), Rosemary (8), Lavender (6), scented Pelargoniums.

History/description: A graphic artist from a farming family, Jeanne Dunn has reverted to her roots. She quit the arts to work in a nursery where she developed a passion for herbs. Wanting to go into business for themselves, this couple opened the nursery in Encinitas in 1983, moving it to Fallbrook in 1990 to gain more space. They have a display garden, give classes, put on herb festivals and can help you plan your own herb garden. They recommend you visit the wonderful herb garden at the Wild Animal Park in Escondido.

How to get there: From Highway 15, take Rainbow Valley Boulevard east to old Highway 395. Go south to 2nd Street. Go east on 2nd Street.

General information: Open everyday, 9 am - 4 pm. They plan to make shipping available soon. Plant list available for S.A.S.E.

County: San Diego

JUDY'S PERENNIALS
436 Buena Creek Road
San Marcos 92069
(619) 471-8850
Judy Wigand

Retail

Plant Specialties: Perennials, including Salvia (15 varieties), Penstemon (10), Alstroemeria (Meyer hybrids), Australian and Mediterranean plants. Although she specializes in perennials suited to the south and central coast, over half her stock is cold hardy.

History/description: This self-taught, slightly mad perennial grower, raised her kids, shed her apron and went outdoors to garden. For five years she worked in the daylily nursery next door, while tending her garden at home. Customers of the nursery, noticing her well-grown and unusual plants, wanted to buy. Business grew as well as her plants; she now oversees a half acre of stock and a display garden. She will give talks to groups.

How to get there: From Highway 5 at Oceanside, go east on Highway 78. Continue for 15 miles to San Marcos and exit at Twin Oaks Valley Road. Go north on Twin Oaks Valley Road for 3 miles. Go west on Buena Creek Road for 1/2 mile.

General information: Open Wednesday through Friday, 10 am - 4 pm. First and third Saturdays.

County: San Diego

KARTUZ GREENHOUSES

1408 Sunset Drive
Vista 92083-6531
(619) 941-3613
Michael Kartuz

Retail

Plant Specialties: Flowering subtropical plants, large collection of Begonias (cane-type, rex, rhizomatous, shrubby), flowering Vines, including Passiflora, Hoya. Indoor plants.

History/description: In 1979, Michael Kartuz brought his business west from Wilmington, Delaware, attracted by the infinitely nicer weather and lower heating costs. Although most plants are still grown in greenhouses, they summer (April to October) outside in shade houses. He will custom grow for clients.

How to get there: Call for directions.

General information: Open Wednesday through Saturday, 9 am - 4 pm. Best to call first. They will ship mail orders. Catalogue available for $2.

County: San Diego

MAGIC GROWERS

1156 E. Green Street (mailing address)
Pasadena 91107
(818) 797-6511
Lance Tibbet, Joe Brosius, Bert Tibbet

Wholesale

Plant Specialties: California native plants, Australian plants, Mediterranean plants, Herbs, Perennials, Drought tolerant plants.

History/description: Noticing that there were no growers of perennial plants in the Pasadena area in the late 1970s, Bert Tibbet saw an opportunity. Originally intending to open a retail nursery but running afoul of the city permit process, he and his horticulturist partner, Joe Brosius, launched a wholesale business instead. Son Lance has now taken over supervision of their huge stock selection and display garden.

How to get there: Call for directions. (Address listed is mailing address only.)

General information: Open to wholesale trade only, Monday through Friday, 7:30 am - 4:30 pm. Retail customers may visit with a landscaper. They will ship plants within southern California. Plant list is available for S.A.S.E. Tours of the nursery can be arranged and staff is available for lectures.

County: Los Angeles

NORTHRIDGE GARDENS

9821 White Oak Avenue
Northridge 91325
(818) 349-9798
Arnie Mitchnick

Retail and Wholesale

Plant Specialties: Succulents, including Caudiciforms, "bonsaiable" succulents, Sansevieria. Trees, including Ficus, Brachychiton, Chorisia, all unusual and hard-to-obtain.

History/description: A hobbyist for many years, Arnie Mitchnick regularly bought plants from Manny and Bert Singer of Singer's Growing Things. When Singer's wanted to scale down their business, Arnie bought some of their incredible collection. Suddenly this writer/editor and former chemistry and biology teacher found himself in the nursery business. He is presently expanding his display gardens and gives tours of his premises to groups. Plan also to visit the Huntington Library in Pasadena and Descanso Gardens in Altadena.

How to get there: From Highway 118 (Simi Valley Freeway), take Balboa exit. Go south on Balboa to Lassen. Go right on Lassen. Go left on White Oak (It becomes a dirt road). From Highway 101, take Balboa exit north, then left on Lassen.

General information: Open Friday and Saturday, 9 am - 5 pm or by appointment. They ship plants. Their catalogue is free.

County: Los Angeles

NUCCIO'S NURSERIES
3555 Chaney Trail,
P. O. Box 6160, Altadena 91003
Altadena 91001
(818) 794-3383
Tom, Jim and Julius (Jr.) Nuccio

Retail and Wholesale

Plant Specialties: Wide selection of species Camellias. Special hybrids include 'Giulio Nuccio', 'Nuccio's Cameo'. 'Nuccio's Carrousel', 'Spring Festival'. Over 700 varieties of Azaleas, with more than 100 varieties of Satsuki azaleas.

History/description: Brothers Joe and Julius Nuccio started this nursery in 1935. Their sons now run the business, just as fascinated by the scientific possibilities of hybridizing and as thristy for new introductions as were their parents. Every year they introduce three or four new varieties from China and Japan (20 in 1992). They work on staying small and keeping focused on exceptional plants. A display garden educates visitors and offers a major springtime spectacle.

How to get there: Take the Ventura Freeway (Highway 134) into Pasadena, It become the Foothill Freeway (Highway 210). Go north off of 210 onto Fair Oaks Avenue. After six miles, go west on Loma Alta. Go north on Chaney Trail. Nursery is at end of road.

General information: Open Friday through Tuesday, 8 am - 4:30 pm. They ship plants all over the world. Free catalogue is available.

County: Los Angeles

PECOFF BROS. NURSERY AND SEED

20220 Elfin Forest Road
Escondido 92029
(619) 744-3120
Ronald Pecoff

Retail and Wholesale

Plant Specialties: Over 350 species of Trees, Shrubs, Ground Covers and their Seeds. Specializing in plants grown for use in adverse conditions - dunes, cut and fill slopes, erosion/bank stabilization, salinity, drought and fire. Though stock is constantly changing, he often carries Acacia 'Pecoff Verde', Acacia 'Onderup', Atriplex glauca, Galvesia speciosa, as well as many other natives and exotics.

History/description: Horticulturist Ronald Pecoff started this business in 1963, selling native plants and seeds exclusively. He developed a hydro-seeding technology which enabled disturbed areas to be planted with plants as well as with seeds. He specializes in contract growing for large-scale projects; the remainder sold at his retail nursery accounts for only 10% of his business. He has worked on sand stabilization projects in Qatar, Iran, and Dubai and is presently restoring a mine tailings site in New Guinea.

How to get there: Call for directions.

General information: Open Monday through Friday, 6:30 am - 5 pm. Call first. He will ship native plant and exotic seed.

County: San Diego

RAINBOW GARDENS NURSERY AND BOOKSHOP

1444 East Taylor Street
Vista 92084
(619) 758-4290
Chuck Everson

Retail

Plant Specialties: Tropical Cacti from the jungle, including Epiphyllum (200 varieties), Rhipsalis, Hoya, Schlumbergia.

History/description: Who knows how you develop jungle fever? Chuck Everson caught the bug when a business opportunity presented itself. In 1980 he bought out a nursery and merged it with his existing bookstore which specialized in books about horticulture and cactus culture. He claims he now has every book in print about bromeliads, epiphyllums, palms and succulents to help you make an educated purchase.

How to get there: Call for directions.

General information: Open April and May, Tuesday through Saturday, 9:30 am - 4 pm. Rest of year, by appointment. 28 page catalogue with color photos is available for $2.

County: San Diego

RAINFOREST FLORA

1927 West Rosecrans
Gardena 90249
(310) 515-5200
Jerry Robinson; Paul Iseley, President

Retail and Wholesale

Plant Specialties: Tillandsia, Staghorn Ferns, Cycads, Bromeliads, Neoregelia.

History/description: That little Tillandsia plant glued on wood that you picked up at Home Depot probably got its start here. Although their business is now big and mostly wholesale, they started out small. In 1974 these two Bromeliad collectors went to Mexico and Guatemala to explore and collect and the partnership was born. They now have 7 greenhouses on 2 acres plus additional growing grounds in San Diego. They maintain a display garden and are the only nursery which grows its own stock of Tillandsia.

How to get there: From Los Angeles, go south on Highway 405 for 3 miles to Rosecrans. Go east on Rosecrans for 3.5 miles past Van Ness Avenue. Their driveway is between Nissin Foods and the Southern California Edison substation.

General information: Open Monday through Saturday, 8 am - 4:30 pm. They ship plants. Plant list is avialable for S.A.S.E. Catalogue costs $2.

County: Los Angeles

RICHARD STRETZ, HYBRIDIZER

1929 Rome Drive
Los Angeles 90065
(213) 221-6167
Richard Stretz

Retail

Plant Specialties: Daylilies, 100 varieties, all hybrids, developing re-bloomers for three season color. Reblooming Iris.

History/description: This Missouri farmboy took his time getting back to earth. He studied acting with Lee Strasberg and Ute Hagen in New York and appeared in films and on stage, including a stint in "spaghetti westerns" in Italy. Back in the US, he got a degree in agriculture and floriculture and started designing for estate gardens in Los Angeles. When not working on his daylilies, he is in arts education for the city of Los Angeles. He has a display garden, can put together a slide show or talk and will design a perfect spot for your daylilies. He recommends you visit Barnsdell Park, 4800 Hollywood Boulevard, an art park with two Frank Lloyd Wright houses.

How to get there: Call for directions.

General information: Open by appointment only. He will ship plants. Mail order catalogue is available.

County: Los Angeles

SAN GABRIEL NURSERY

632 South San Gabriel Boulevard
San Gabriel 91776
(818) 286-0787
Yoshimura Family; Herick Mar, Manager

Retail and Wholesale

Plant Specialties: Shrubs, Bulbs, Aquatic plants, Bonsai, large selection Camellia japonica, Asiatic Shrubs, including Michelia 'Alba' and M. champaca. Hallmark is variety.

History/description: Founded by Fred Yoshimura in 1923, his family has been involved ever since. After internment during World War II, the Yoshimura family was able to buy back the nursery in 1946. Everyone's recommendation for an all-around nursery in Los Angeles, they grow most of their own plants. There is a florist shop on the premises.

How to get there: Thirteen miles east of Los Angeles on Highway 10, take San Gabriel Boulevard exit. Go two miles north.

General information: Open everyday, 8 am - 5 pm.

County: Los Angeles

SARVER NURSERY
2600 Sarver Lane
San Marcos 92069
(619) 744-0600
Rosalind Sarver

Retail and Wholesale

Plant Specialties: Azaleas (75 varieties), including Belgian and Southern Indica. Canna (40 varieties, large-bulbed).

History/description: Mrs. Sarver and her late husband have been plantspeople from way back. They had an azalea nursery in Dallas where they perfected their peat-grown technique. They moved west to retire in 1954 and figured they might as well repeat their success. Cannas were introduced in 1958 when they discovered how well they grew. Quite simply, she is called the sweetheart of San Diego county. She still gives tours of the nursery and its 200 year old oak trees. Sarvers will custom grow.

How to get there: Take Highway 5 to Oceanside exit. Go west toward Highway 15, past Vista and San Marcos. Go left on Twin Oaks Valley Road. In four miles the road forks, go right on Deer Spring Road for one crooked mile. Sarver Lane is on the north side of the road.

General information: Primarily wholesale, but self-service retail customers admitted, Monday through Saturday, 7 am - 4 pm. All customers must call first. They ship cannas only. Plant list available for S.A.S.E.

County: San Diego

SERRA GARDENS

3314 Serra Road
Malibu 90265
(310) 456-1572
Don Newcomer

Retail and Wholesale

Plant Specialties: Cacti and succulents, from all parts of the world, including Pachycormis discolor, Pedilanthus macrocarpus, Geradanthus macrorhizus, Ruschia lineolata, Aloe thraskii, A. bainsii, Agave, Sedum, Euphorbia. Large plants for landscape use to 2" pots.

History/description: In 1986 Don Newcomer made a business out of his father's 30 year old cacti and succulent habit. His well-established garden is on his father's 10 acre property nestled in a Malibu canyon. Serra Gardens propagates by tissue culture and seek to preserve species in endangered habitats. They contract grow and offer design services. Plan to visit the nearby gardens of the Getty Museum and Michael Landon Community Center.

How to get there: Call for directions.

General information: Open by appointment only. They ship clones only. Mail order catalogue available for S.A.S.E.

County: Los Angeles

STALLINGS RANCH NURSERY

910 Encinitas Boulevard
Encinitas 92024
(619) 753-3079
Jack Porter; Dale Kolaczkowski, Manager

Retail

Plant Specialties: Unusual Subtropical plants, including 150 Hibiscus cultivars, 16 Gingers, 12 Thunbergia, Jasmine, Vines. Some hardy plants.

History/description: In 1978 Jack Porter bought Virgel Stallings' all-purpose nursery which Mr. Stallings opened just after the war in 1945. The emphasis is now on unusual and hard-to-find plants and on their special collections. They offer site-specific consultations in their local area. Plan also to visit Quail Botanical Gardens and Balboa Park.

How to get there: Call for directions.

General information: Open Monday through Saturday, 8:30 am - 5 pm; Sunday, 8:30 - 4 pm. Winter Hours: Monday through Saturday, 8:30 am - 4:30 pm. Closed Christmas week and major holidays. They will ship plants. Catalogue available for $3, refundable with purchase.

County: San Diego

THEODORE PAYNE FOUNDATION

10459 Tuxford Street
Sun Valley 91352
(818) 768-1802
Dennis Bryson, Sales Manager

Retail and Wholesale

Plant Specialties: 700 species of California native plants and seeds, (100 of which are endangered), large collections of Arctostaphylos, Ceanothus. Mahonia nevenii saved from extinction.

History/description: Named in honor of Theodore Payne (1872-1963), this non-profit foundation was established in 1961 to preserve native vegetation through propagation and to teach the public about its value. Theodore Payne's impact on the California landscape was immense, occuring at a critical time when underappreciated native vegetation was being lost to farming and progress. As a nurseryman and designer, his hand helped shape Rancho Santa Ana Botanic Garden, Los Angeles State and County Botanic Gardens and Torrey Pines State Park. He popularized California natives in his native England, even before they caught on at home. The Foundation has classes, field trips, library, museum, horticultural consultants, and a Wildflower Hotline (818) 768-3533. Their 21 acre site includes a 5 acre wildflower hill, trails, and picnic facilities.

How to get there: From Highway 210 (Foothill Freeway), take La Tuna Canyon exit. Go west for four miles, then right at signal onto Wheatland. Go right on Tuxford. Nursery in half block on leFort

General information: Open Tuesday through Saturday, 8:30 am - 4:30 pm. They ship books and seeds only. Catalogue and plant list available for $2.

County: Los Angeles

TREE OF LIFE NURSERY

33201 Ortega Highway,
P. O. Box 736
San Juan Capistrano 92693
(714) 728-0685, FAX (714) 728-0509
Mike Evans, Jeff Bohn

Retail and Wholesale

Plant Specialties: California native plants for all regions of the state, especially plants of the Coastal mountains.

History/description: These former landscape contractors and native plant enthusiasts knew from their garden installation experience that California natives made first rate garden plants. They gambled that a nursery could be successful, devoted exclusively to the propagation of natives, and opened Tree of Life in 1981. Presently 20 of its 40 acres are under cultivation, including a display garden. They are very involved with ecological landscaping and revegetation work.

How to get there: Take Ortega Highway exit off of Highway 5. Go east for 7.75 miles. Nursery is on the leFort

General information: Open by appointment only. Catalogue is available for $3.50.

County: Orange

TROPIC WORLD

26437 North Centre City Parkway
Escondido 92026
(619) 746-6108
Paul Hutchison

Retail

Plant Specialties: Mainly Cacti and Succulents, including 100 types of Agave and agavaceae, Aloe, Crassula ovata 'Christmas Cheer' (his hybrid). Also 500 varieties of Roses, landscape Shrubs and Trees, including tropical fruit trees (especially Brazilian guavas), Australian and South African plants, Bamboo, Palms, Perennials, Mexican and Texan Salvias.

History/description: Paul Hutchison has had a lifelong fascination with plants. As a teenager working for the Hummel family in summer, he was profoundly affected by Hummel's pioneering interest in cacti and succulents. Trained as a botanist at UC, he taught at Berkeley, was a senior botanist at UC Botanical Garden and worked at Lotusland. Supremely prepared, he opened Tropic World in 1968 and continues to devote himself to the study, collection and hybridization of plants. The nursery is also the site of Tropic World Foundation, established to promote botany and horticulture. An on-site library, a large display garden, and an 11,000 square foot greenhouse containing a major part of the Hummel collection (euphorbia, bromeliads, orthophylum) make learning a pleasure. (While near S.D. plan to visit Quail Botanical Gardens, and the garden at the Bahia Resort Hotel.)

How to get there: From Highway 15, go east on Deer Springs Road to North Centre City Parkway. Go south for half mile.

General information: Open everyday, 9 am - 5 pm. He is considering shipping plants. Check with him.

County: San Diego

VAN NESS WATER GARDENS

2460 North Euclid Avenue
Upland 91786
(909) 982-2425
Bill Uber

Retail

Plant Specialties: Aquatic plants and aquatic ecosystem components, bog plants, small flowering ornamentals appropriate to damp places surrounding water gardens.

History/description: There have been a lot of changes in Upland since this family business was launched in 1922. The lemon grove which surrounded the nursery then is today a residential area. Three generations have run the business; the original Mr. Van Ness was a cousin of Bill Uber's father. Bill's father took over in 1952. It was Bill's turn in 1976. They must have the kinks out of their hoses by now. They have a display garden, give tours of the site and will help you plan your water garden.

How to get there: From Highway 10, take Euclid exit in Upland. Go 4-1/2 miles north. When the double lane ends, Euclid forks; take the left fork for 1/4 mile to base of Mt. Baldy. Nursery is on the right.

General information: Open Tuesday through Saturday, 9 am - 4 pm. They will ship mail orders. Catalogue is available for $3.

County: San Bernardino

WILDWOOD NURSERY

3975 Emerald Avenue,
P. O. Box 1334, Claremont 91711
LaVerne 91750
(714) 593-4093
Bonnie Walsh

Retail and Wholesale

Plant Specialties: California native plants, Drought tolerant plants, Mediterranean climate plants. Large collection of Ceanothus, heat tolerant Perennials, Artemisia (species and cultivars). Wild Seed. Revegetation plant material for wholesale only.

History/description: Wildwood Nursery will continue to supply plant lovers with fascinating fare despite the untimely death of Ray Walsh, landscape architect, revegetation specialist, and good friend of California horticulturists. Bonnie Walsh has the right combination of enthusiasm, energy, and grit to make the nursery thrive. In business since 1983, her 1-1/3 acre site is in an old orange grove, just west of Claremont. The nursery will contract grow and will collect seed on contract. While in the area, plan to visit the California native garden near Highway 605 in Baldwin Park and take the 42 mile scenic drive along Glendora Mountain Road, Glendora Ridge Road and Mt. Baldy Road.

How to get there: From Los Angeles, take Highway 210 east to Highway 30 east onto Foothill Boulevard. At Emerald go left (north) toward mountains.

General information: Call for open hours. They will ship plants. Catalogue available for $1.

County: Los Angeles

HORTICULTURAL ATTRACTIONS

Descanso Gardens

1418 Descanso Drive
La Canada Flintridge 91011
(818) 952-4400, (818) 790-5414 (Guild)

History/description: Manchester Boddy lived the good life of a Los Angeles newspaper publisher in a 22 room house he built for himself in 1937 on this 165 acre piece of the old Rancho San Rafael. Although bought by the county in 1954, the gardens' future was secure only when a group of concerned neighbors formed the Gardens Guild, a membership support group. Their efforts got the property placed under the jurisdiction of the LA County Department of Arboreta. Boddy's Rose and Camellia Gardens remain highlights in this oak-woodland canyon setting. Also featured are a Japanese garden, a Lake Bird Sanctuary, Fern Canyon, Iris Garden and Native Plant Garden. Plant Societies regularly hold shows here and the guild offers lectures, Christmas Show, docent tours and events. Admission is $3; free on third Tuesday of each month.

General information: Garden open everyday 9 am - 5 pm. Closed Christmas. Gift Shop open 9 am - 4:30 pm.

Plant sales: Big plant sales are held in April and October. Plants also sold at the Gift Shop.

County: Los Angeles

Fullerton Arboretum

California State University,
P. O. Box 34080
Fullerton 92634-9480
(714) 773-3579 (Friends)

History/description: This 26 acre arboretum was opened in 1979 through the joint effort of the California State University at Fullerton, the City of Fullerton, and community resources. The arboretum includes plants from all parts of the world which can grow well in southern California and organizes them according to their cultural requirements. Hence, moisture-loving plants grow alongside their stream and ponds, and dryland plants are found in the chaparral section. Other areas include Palm Gardens, Subtropical Fruit Grove, Conifer Area, Cactus and Succulent Garden, and the Historic Area, a re-creation of an 1890s garden surrounding a Victorian house. The Friends, a membership support group, organizes events, newsletter, and the shop.

Plant sales: Plants are available at the Garden Shop and at plant sales held October through June, on Saturdays, 10 am - 4 pm.

General information: Arboretum is open everyday, 8 am - 4:45 pm, except Thanksgiving, Christmas and New Year's. Library is open Thursday, noon - 3 pm. Garden Shop is open Tuesday through Saturday, 11 am - 2 pm; Sunday, 1 pm - 4 pm, closed in August.

County: Orange

Huntington Library Art Collection & Botanical Gardens

1151 Oxford Road
San Marino 91108
(818) 405-2282 (Gardens), (818) 405-2100 (Main)

History/description: Moving his uncle's railroad

and real estate empire south, Henry Huntington bought the San Marino Ranch in 1904. He spent the rest of his life creating this estate to house his art and book collections. In 1919 Henry and his wife Arabella deeded the property to a non-profit trust to insure its perpetuity. His pleasure garden is today's 130 acre botanical garden. Fifteen specialized gardens include a Desert Garden, the Japanese Garden, Rose Garden and Camellia collection. Primarily a display garden, the Huntington also has an herbarium, mounts plant expeditions and issues botanical publications. Entrance is free but a $5 donation is suggested. The Huntington's many services include docent tours, classes, workshops and lectures.

General information: Open Tuesday through Friday, 1 pm - 4:30 pm; Saturday and Sunday, 10:30 am - 4:30 pm.

Plant sales: Plants are sold on the first Thursday of each month during the Garden Talk and at their annual plant sale in May.

County: Los Angeles

Los Angeles State and County Arboretum

301 North Baldwin Avenue
Arcadia 91007-2697
(818) 821-3222

History/description: The Arboretum is the cornerstone of the consortia of horticultural attractions administered by the Department of Arboreta and Botanic Gardens (namely Descanso Gardens, South Coast Botanic Garden and Virginia Robinson Gardens). Its 127 acres include more than 4,000 species of plants from around the world. The grounds include several historical buildings and a 3.5 acre lake. The faint-footed can get around by tram ($4.50). Entry for self-guided walkers is $3. Sights include an Herb Garden, Tropical Greenhouse,

Demonstration Garden, Jungle Garden, Water Conservation Garden. Founded in 1947, the arboretum has introduced more than 100 plants for southern California landscapes. The California Arboretum Foundation, a membership support group, organizes trips, lectures, events, and can get you a discount on purchases at participating nurseries.

General information: Arboretum is open everyday but Christmas, 9 am - 5 pm. Gift Shop is open from 10 am - 4 pm.

Plant sales: Their big plant sale, the Baldwin Bonanza, takes place in early May. Plants are also available at the many plant society shows held at the arboretum. A few are sold at the Gift Shop.

County: Los Angeles

Quail Botanical Gardens

230 Quail Gardens Drive
Encinitas 92024
(619) 436-3036

History/description: In 1957 Ruth Baird Larabee donated her home and 26 acre garden collection of drought tolerant plants to the county. An additional four acres were given by Poinsettia grower Paul Ecke in 1971, The Foundation was formed in 1961 to provide support for the garden and enable it to grow botanically. The Gardens are known throughout the country for having the most diverse collection of Bamboo; they maintain a bamboo quarantine house for imports. Other special collections include Native Plants, Pan-tropical plants, Hibiscus, Palms, Cycads. Also there is a Fruit Demonstration Garden, Desert Garden, Overlook Pavilion and Sub-tropical waterfall. Docent tours are given every Saturday at 10 am; children's tours are on the first Tuesday of each month.

General information: Open everyday except

Monday, 8 am - 5 pm. Nursery open Wednesday - Sunday, 11 am - 3 pm.

Plant sales: Plants sold at the nursery.

County: San Diego

Rancho Santa Ana Botanic Garden

1500 North College Avenue
Claremont 91711
(909) 625-8767

History/description: Rancho Santa Ana was the first botanical garden in California devoted to scientific study of its native flora. Founded in the 1920s on the Orange County ranch of its benefactor, Susanna Bixby Bryant, the garden served as a catalyst for the efforts of native plant pioneers. In its present location since 1951 and affiliated with the Claremont Colleges, the garden offers certificate courses, and public classes. A leading research center, it has an herbarium and library. Half of the garden's 87 acres are devoted to research projects; half is public, organized by plant communities - desert, coastal, woodland, riparian, with special manzanita, ceanothus, conifer and wildflower display areas. A new California Cultivar Garden highlights cultivated varieties of native flora. The Friends, a membership support group, offers guided tours, March through May, Saturday at 9 am, Sunday at 2 pm.

General information: Open everyday, 8 am - 5 pm. Closed Christmas, New Year's, July 4 and Thanksgiving. The gift shop is open weekdays, 9 am - 4:30 pm; weekends, 11 am - 4 pm.

Plant sales: Plants are sold at the gift shop and at their big plant sale the first weekend in November (7000 plants for sale).

County: Los Angeles

South Coast Botanic Garden

26300 Crenshaw Boulevard
Palos Verdes Peninsula 90274
(310) 544-1847

History/description: The South Coast Botanic Garden is California's first major successful reclamation project. Formerly a landfill, its natural setting gives no clue to its past. Specialized gardens on its 87 acres include a Waterwise Garden, Rose, Cactus, Fuchsia and Palm Gardens, Pine, Ficus and Flower areas, and a Garden for the Senses. Admission is $3. Everything from weddings and concerts to flower shows takes place here.

General information: Garden is open everday except Christmas, 9 am - 5 pm.

Plant sales: The Foundation, a membership support group, operates a gift shop which sells plants, as well as food for the resident ducks. Big plants sales take place in the fall and on the third weekend in May (Fiesta de Flores).

County: Los Angeles

UC Irvine Arboretum

University of California
Irvine 92717
(714) 856-5833

History/description: Specializing in plants of the five Mediterranean ecosystems, the Arboretum began as the holding area for plants used to landscape the campus in 1964. Noted for its collection of South African bulbs and African Aloes, the Arboretum is involved in the conservation of endangered species and maintains a gene bank. The Friends, a membership support group, publishes a quarterly newsletter, gives lectures and supports conservation activities.

General information: Open Monday through

Saturday, 9 am - 3 pm. Closed University holidays.

Plant sales: Large bulb and plant sale is held in August. Smaller sales accompany monthly events.

County: Orange

Other Attractions:

La Casita de Arroyo
177 South Arroyo Boulevard
Pasadena 91105

Original garden (1932) redesigned by Isabelle Greene.

Hannah Carter Japanese Garden
University of California, Los Angeles
10619 Bellagio Road
Los Angeles 90024
(310) 825-4574

Bel Air garden of Gordon Guiberson, authentic Japanese details.

Chavez Ravine Arboretum
Elysian Park
Los Angeles
(213) 485-5054

Survivors of Los Angeles' first arboretum.

Gamble House
4 Westmoreland Place
Pasadena 91103
(818) 793-3334

Early California garden with Greene & Greene designed house.

J. Paul Getty Museum Gardens
17985 Pacific Coast Highway
Malibu 90265
(310) 458-2003

Re-creation of Roman garden with Main Peristyle, Inner Peristyle, and Herb Gardens. Plantings based on historical records as much as possible.

Charles F. Lummis House, "El Alisal"

200 East Avenue 43
Highland Park area of Los Angeles, 90031
(213) 222-0546

Almost two acre garden redesigned in 1986 by Robert Perry using water-wise plantings and gardening techniques. Current home of Historical Society of Southern California.

Mildred E. Mathias Botanical Garden

University of California, SW corner of campus
Los Angeles, 90024
(310) 825-3620

Rare trees, landscaped displays of mediterranean and subtropical plants.

"El Molino Viejo", Old Mill

1120 Old Mill Road
San Marino 91108
(818) 449-5450

Walled mission garden with old citrus and fruit trees and Spanish tub-mill. Pomegranate patio and plantings display plants used by the padres.

Franklin Murphy Sculpture Garden

University of California, NE corner of campus

Los Angeles 90024
(310) 825-9345

Sculpture garden with trees and earthwork, designed by Ralph D. Cornell.

Pomona College

333 N. College Way
Claremont 91711
(714) 621-8146

Collegiate landscape and small gardens by Ralph D. Cornell.

Rancho Los Alamitos

6400 Bixby Hill Road
Long Beach 90815
(310) 431-3541

Adobe with garden of the 1920s designed by Florence Yoch and others.

Sherman Library and Gardens

2647 E. Pacific Coast Highway
Corona Del Mar 92625
(714) 673-2261

Horticultural display garden, with rose garden, tropical greenhouse, large annual flower beds and hanging basket displays.

Torrey Pines State Reserve

One mile south of Del Mar on N. Torrey Pines Road
(619) 755-2063

The only remaining Torrey Pine forest, on 1,100 acres of beach, bluff, and coastal mountain terrain.

UC Riverside Botanical Garden

University of California
Riverside 92521
(714) 787-4650

Desert and tropical plants displayed over rugged terrain.

Virginia Robinson Gardens

1008 Elden Way
Beverly Hills 90210
(310) 276-5367

Terraced estate garden with palm grove, citrus garden, exotic trees.

Wisteria Vine

Sierra Madre 91021
(818) 355-5111 (Chamber of Commerce)

World's largest wisteria vine gone rampant over two private backyards; call for notification of peak bloom visiting days in March.

Wrigley Memorial Botanic Garden

1400 Avalon Canyon Road
Avalon, Catalina Island, 90704
(310) 510-2288

Cactus and succulent garden; California native plant garden, emphasizing flora of the Channel Islands.

Other Sources of Plants

OREGON & WASHINGTON GROWERS

When you find yourself far afield, visit these special growers in Oregon and Washington. Always call before you go, as hours are irregular. Most do mail order.

COLLECTOR'S NURSERY
Diane Reeck, Bill Janssen
1602 NE 162nd Avenue
Vancouver, OR 98684
(206) 256-8533
Plants: Unusual plants, Dwarfs, Alpines, rare Perennials. Catalogue available.

BALTZER'S SPECIALTY PLANTS
Bob Baltzer
36011 Highway 58
Pleasant Hill, OR 97455
(503) 747-5604
Plants: Rare Conifers, Maples and other deciduous Trees.

BLOOMING NURSERY
Grace Dinsdale
3839 SW Golf Course Road
Cornelius, OR 97113
(503) 357-2904
Plants: 3 acres of Perennials in bedding format.

CAPRICE FARMS NURSERY
Dot and Al Rogers
15425 SW Pleasant Hill Road
Sherwood, OR 97140
(503) 625-7241
Plants: Peonies, Daylilies, Hostas, Japanese Iris, Siberian Iris. Catalogue available.

FORESTFARM
Peg and Ray Prag
990 Tetherow Road
Williams, OR 97544
(503) 846-6963
Plants: Western native plants, hard-to-find Perennials, Trees and Shrubs, hardy Eucalyptus, Fruiting plants, Bee plants, Conifers, Dye plants. Catalogue available.

GOODWIN CREEK'S SECRET GARDEN
Dottie and Jim Becker
154 1/2 Oak Street
Ashland, OR 97520
(503) 488-3308
Plants: 350 varieties of Herbs. Everlasting Perennials, Seeds. Also book, *Concise Guide to Growing Everlastings*. Catalogue $1.

GOSSLER FARMS NURSERY
Marge and Roger Gossler
1200 Weaver Road
Springfield, OR 97478
(503) 746-3922
Plants: Magnolias, very unusual Shrubs and Trees. Catalogue $1.

GREER GARDEN
Harold Greer
1280 Goodpasture Island Road
Eugene, OR 97401
(503) 686-8266
Plants: Rhododendrons, rare and unusual Shrubs and Trees. Catalogue $3.

NICHOLS NURSERY
Rosemary Nichols-McGee
1190 North Pacific Highway
Albany, OR 97321
(503) 928-9280
Plants: Culinary Herbs. Herb, flower and vegetable Seeds. Plants sold after April 15. Catalogue $1. Festive open house in May.

PORTERHOWES FARMS
Don Howes, Lloyd Porter
41370 SE Thomas Road
Sandy, OR 97055
(503) 668-5834
Plants: Alpines, Rock Garden plants, rare and Dwarf Conifers, unique Broad-leaved Trees and Shrubs.

RED'S ALPINE GARDENS
Dick Cavender
15920 SW Oberst Lane
Sherwood, OR 97140
(503) 625-6331
Plants: 400 species and hybrid Rhododendrons and Azaleas. Catalogue available.

SISKIYOU RARE PLANTS
Baldassare Mineo
2825 Cummings Road
Medford, OR 97501
(503) 236-8024
Plants: Alpine plants, Rock Garden plants. Old and well-respected nursery.

THE BOREES NURSERY
Sorenson and Watson
1737 SW Coronado
Portland, OR 97219
(503) 244-9341
Plants: Rhododendrons, Azaleas, extensive collection of companion plants, deciduous Trees, Alpines and Rock Garden plants, Woodland plants.

BARFOD'S HARDY FERNS
Torben Barfod
23622 Bothell Way
Bothell, OR 98021
(206) 483-0205
Plants: Ferns, waterfall display beds.

FANCY FRONDS
Judith Jones
1911 4th Avenue West
Seattle, OR 98119
(206) 284-5332
Plants: Ferns. Catalogue $1, refundable.

GRAND RIDGE NURSERY AND POTTERY
Phil Pearson, Steve Doonan
27801 SE High Point Road
Issaquah, OR 98027
(206) 222-7226
Plants: Rock Garden plants.

HERONSWOOD NURSERY
Don Hinckley
7530 288th Avenue NE
Kingston, OR 98346
(206) 297-4172
Plants: Perennials, Shrubs, Shade plants.

MT. TAHOMA NURSERY
Rick Lupp
28111 112th Avenue East
Graham, WA 98338
(206) 847-9827
Plants: Rock Garden plants, Woodland plants.

ROBYN'S NEST NURSERY
Robyn Duback
7802 NE 63rd Street
Vancouver, WA 98662
(206) 256-7399
Plants: Shade plants, including Hostas and 35 varieties of Astilbe. Ornamental Grasses. Catalogue $2. Open May through June, September and October.

SEEDS & BULBS

For those who like to start small, here are sources for seeds and bulbs. Many of these plants are only available from seed. Practically all listings are mail order only.

ALBRIGHT SEED COMPANY
487 Dawson Drive
Camarillo 93012
(800) 423-8112

Plant Specialties: California native wildflowers seeds and grass seeds, legume seeds, pasture grass seeds, turf grass seeds for erosion control, revegetation and cover crops. Hand-harvested and custom-harvested seeds. Free catalogue available.

BOUNTIFUL GARDENS
18001 Shafer Ranch Road
Willitts 95490
(707) 459-6410

Retail

Plant Specialties: Untreated, open-pollinated seeds for vegetables, grains, cover crops, herbs, and some flowers. Primarily heirloom plants. Also books and organic gardening supplies. Operated by non-profit Ecology Action in Palo Alto. Free catalogue. Mailorder only.

CARTER SEEDS
475 Mar Vista Drive
Vista 92083
(619) 724-5931

Retail and Wholesale

Plant Specialties: Seeds: flowers, trees, shrubs, ornamental grasses, palms. Catalogue is free.

CHOICE EDIBLES
584 Riverside Park Road
Carlotta 95528
(707) 768-3135

Retail and Wholesale

Plant Specialties: Shiitake mushrooms, sawdust-block kit ready to sprout. Other mushrooms available for wholesale only. Free flyer.

CLYDE ROBBIN SEED COMPANY
3670 Enterprise Avenue
Hayward 94545
(510) 785-0425

Retail and Wholesale

Plant Specialties: Primarily U.S. native wildflower seeds, grass seed and legume seed. shrub seed and tree seed. 100% grown, cleaned and combined - on location from Utah to California. 600 kinds of seeds. Free catalogues available for wildflowers and for shrubs and trees.

CONSERVASEED
P. O. Box 455
Rio Vista 94571
(916) 775-1676

Wholesale primarily, some Retail

Plant Specialties: Largest producer of California native grass seed, legume seed and forb seed for erosion control and revegetation. A genetic collector and increaser. Current catalogue is free. There may be a charge for new catalogue.

ENVIRONMENTAL SEED PRODUCERS
P. O. Box 2709
Lompoc 91738
(805) 735-8888

Wholesale

Plant Specialties: Seeds for California native wildflowers, ornamental grasses, wildflower mixes, herbs. Free catalogue.

FOREST SEEDS OF CALIFORNIA
1100 Indian Hill Road
Placerville 95667
(916) 621-1551

Retail and Wholesale

Plant Specialties: Forest tree and shrub seeds primarily. Also do custom seed collecting and cleaning. Free catalogue.

GREENLADY GARDENS/SKITTONE BULB CO.
1415 Eucalyptus
San Francisco 94132
(415) 753-3332

Retail and Wholesale

Plant Specialties: 500 uncommon kinds of bulbs and bulbous plants, specializing in bulbs from South Africa but including bulbs from many parts of the world. Terrestrial bulbous orchids. Catalogue available for $3.

HEIRLOOM GARDEN SEEDS
P. O. Box 138
Guerneville 95446
(707) 869-0967

Retail and Wholesale

Plant Specialties: Herb seeds, flower seeds - over 400 varieties of open-pollinated (non-hybrid) seeds. Retail catalogue available for $2.50.

J. L. HUDSON, SEEDSMAN
P. O. Box 1058
Redwood City 94064
No telephone orders.

Retail

Plant Specialties: Rare seed from around the world. Bulbs, vegetables, cyclamen. Catalogue available for $1.

KITAZAWA SEED COMPANY
1111 Chapman Street
San Jose 95126
(408) 243-1330

■262 Seeds & Bulbs

Retail and Wholesale

Plant Specialties: Primarily Asian vegetable seed, including burpless cucumber and Japanese hybrid spinach which grows year-round in hot or cool climate. All seed guaranteed. Free brochure.

LOCKHART SEEDS
P.O. Box 1361 (Mail Orders)
Stockton 95201
(209) 466-4401

Retail and Wholesale

Plant Specialties: Vegetable seeds, including many types of onions. Some flowers. Retail shop at 3 North Wilson Way in Stockton also sells small garden equipment and other gardening products. Free catalogue.

ORNAMENTAL EDIBLES
3622 Weedin Court
San Jose 95132
(408) 946-7333

Retail

Plant Specialties: International vegetable seeds, herb seeds and edible flower seeds. Catalogue available for $2.

PACIFIC COAST SEED COMPANY
7074 D Commerce Circle
Pleasanton 94588
(510) 463-1188

Wholesale

Plant Specialties: Seeds for turf grasses, native grasses, wildflowers. Free catalogue.

PEACEFUL VALLEY FARM SUPPLY
P. O. Box 2209
125 Spring Hill Boulevard
Grass Valley 95945
(916) 272-4769

Retail

Plant Specialties: Wildflower seeds, vegetable seeds, cover crop seeds. Free catalogue.

RAMSEY SEEDS
205 Stockton Street
Manteca 95336
(209)823-1721

Wholesale

Plant Specialties: Native grasses, clovers, corn, alfalfa, and special mixes. Free catalogue.

REDWOOD CITY SEED COMPANY
P. O. Box 361
Redwood City 94064
(415) 325-7333

Retail and some Wholesale

Plant Specialties: Open-pollinated and old fashioned seed varieties of wildflowers, vegetables, native grasses, shrubs and trees. Many unique to the trade. Retail catalogue available for $1, wholesale list for $2.

ROBINETT BULB FARM
P. O. Box 1306
Sebastopol 95473
(707) 829-2729

Mostly Retail, some Wholesale

Plant Specialties: Bulbs and seeds native to regions west of the Sierra crest, specializing in Alliums, Brodiaea, Calochortus. Most are exclusively theirs. Free catalogue.

S AND S SEEDS
P. O. Box 1275
Carpinteria 93013
(805) 684-0436

Wholesale

Plant Specialties: Seeds for California wildflowers and native grasses. Catalogue available for $6.

SHEPHERD'S GARDEN SEEDS
6119 Highway 9
Felton 95018
(408) 335-5400

Retail and Wholesale

Plant Specialties: Flower seed for cottage, edible, old-fashioned, cutting, fragrant gardens. Perennial and annual seeds in special collections. European vegetable seed, including hot, hot peppers. All seeds tested on-site. Catalogue available for $1.

TERRITORIAL SEED COMPANY
P. O. Box 157
Cottage Grove, Oregon 97424
(503) 942-9547

Retail

Plant Specialties: Vegetable seeds, flower seeds, herb seeds, all tested in their trial gardens. Also, books and gardening supplies. Free catalogue.

FAMILY FARM GUIDES

Groups in various counties print maps and guides to small, family-run farms, dairies, orchards, ranches and nurseries. When requesting a brochure, be sure to enclose a business-sized, self-addressed, stamped envelope.

ALAMEDA COUNTY FARM TRAILS
2300 N. Livermore
Livermore, CA 94550
(510) 373-6966
Specialties: Fruit, goats, bunnies, wine, plants, vegetables

APPLE TRACT
Greater Tehachapi Chamber of Commerce
P. O. Box 401
Tehachapi, CA 93561
(805) 822-4180
Specialties: Fruit

CALAVERAS COUNTY VISITOR GUIDE
Calaveras County Visitors Association
P. O. Box 637
Angel's Camp, CA 95222
(800) 225-3764
Specialties: Fruit, vegetables, birds, ponies, rabbits, herbs

CENTRAL VALLEY HARVEST TRAILS
Stanislaus County Farm Bureau
P. O. Box 3070
Modesto, CA 95353
(209) 522-7278
Specialties: Nurseries, fruit, vegetables, eggs, honey

CIDER PRESS - APPLE HILL GROWERS
El Dorado County Chamber of Commerce
542 Main Street
Placerville, CA 95667
(916) 621-5885
Specialties: Fruit

COASTSIDE HARVEST TRAILS
San Mateo County Farm Bureau
765 Main Street
Half Moon Bay, CA 94019
(415) 726-4485
Specialties: Fruits, vegetables, wines, fish

COUNTY CROSSROADS
Santa Cruz and Santa Clara County Farm Bureaus
141 Monte Vista Avenue
Watsonville, CA 95076
(408) 724-1356
Specialties: Nurseries, fruit, vegetables, wine, dairy

EL DORADO RANCH MARKETING
El Dorado County Chamber of Commerce
542 Main Street
Placerville, CA 95667
(916) 621-5885
Specialties: Nurseries, fruits, vegetables, herbs, horses, sheep, wine

FARMER TO CONSUMER DIRECTORY
Small Farms Department
University of California
Davis, CA 95616
(916) 757-8910
Specialties: Fruit, vegetables, animals

FARMERS MARKETS AND FARMER DIRECT PRODUCTS
Kern County Board of Trade
P. O. Bin 1312
Bakersfield, CA 93302
(805) 861-2367
Specialties: Fruit, vegetables, eggs, Christmas trees

Family Farms

FORTY-NINER FRUIT TRAIL
Placer County Farm Bureau
P. O. Box 317
Newcastle, CA 95658
(916) 663-2929
Specialties: Nurseries, fruit, vegetables, sheep, wine

HARVEST TIME IN BRENTWOOD
P. O. Box O
Brentwood, CA 94513
(510) 672-5115 (Contra Costa County Farm Bureau)
Specialties: Fruits, vegetables, eggs

NAPA COUNTY FARMING TRAILS
Napa County Farm Bureau
4075 Solano Avenue
Napa, CA 94558
(707) 224-5403
Specialties: Nurseries, fruit, vegetables, herbs, eggs, wool

OAK GLEN APPLE GROWERS
Box 1123
Yucaipa, CA 92399
Specialties: Fruit, wine

SONOMA COUNTY FARM TRAILS
Santa Rosa , CA 95406
(707) 586-FARM
Specialties: Nurseries, fruit, vegetables, animals, eggs, herbs, worms.

SUISUN VALLEY HARVEST TRAIL
City of Fairfield, Office of Public Information
1000 Webster Street
Fairfield, CA 94533
(707) 428-7384
Specialties: Fish, fruit, birds, cheese, wine, vegetables

YOSEMITE APPLES TRAILS
Eastern Madera County Chamber of Commerce
P. O. Box 369
Oakhurst, CA 93644
(209) 683-7766
Specialties: Apples

Horticultural Information

SOCIETIES & GROUPS

CALIFORNIA NATIVE PLANT SOCIETY
909 12th Street, #116
Sacramento 95814
(916) 447-2677
State Secretary

History/description: There are 28 local chapters of this venerable organization dedicated to the preservation of native flora, increasing public knowledge about native plants, and monitoring rare and endangered species. All chapters have meetings, field trips, and sales of hard-to-find native plants. They publish a bulletin and *Fremontia* quarterly; local chapters have monthly newsletters.

CALIFORNIA HORTICULTURAL SOCIETY
1847 34th Avenue
San Francisco 94122
(415) 566-5222

History/description: This is the granddaddy of all horticultural societies. Now 700 members strong, they have meetings on the third Monday of each month at 7:30 pm at the California Academy of Sciences. Meetings consist of a slide lecture and an informative show-and-tell by members about their plants. Unusual plants are often available at these meetings. They also have an annual seed exchange, a circulating library, and field trips. With Western Horticultural Society, Southern California Horticultural Society and Strybing Arboretum Society, they formed the Pacific Horticultural Foundation in 1968 which now publishes *Pacific Horticulture*, a quarterly magazine.

ECOLOGY ACTION OF THE MID-PENINSULA
2225 El Camino Real
Palo Alto 94306
(415) 328-6752

History/description: This is a non-profit research and information-sharing group dedicated to bio-intensive agricultural practices - showing that a small plot of land or mini-farm can be, when properly tended, highly productive and sustainable. There are classes and a library at their retail store, Common Ground, in Palo Alto. Their research and training garden in Willits is not open to the public although tours are given in the summer. They also run Bountiful Gardens, a mail order seed and book business.

SARATOGA HORTICULTURAL RESEARCH FOUNDATION
15185 Murphy Avenue
San Martin 95046
(408) 779-3303

History/description: In response to the post-war building boom's demand for trees and shrubs, a group of nurserymen founded Saratoga Horticultural Research Foundation in 1951 to research and develop a successful flora for California. Today the emphasis remains entirely on plant research at their 2.5 acres in San Martin and the 7 acre testing site in Gilroy. During their first forty years, they introduced more than 120 plants into the trade. Member supported activities include lectures, field trips, and two big plant sales each year - the Fall Festival on the first Sat. in October and a Spring Horticultural Fair. Although not open to the public, tours for groups may be arranged.

SOUTHERN CALIFORNIA HORTICULTURAL SOCIETY
P. O. Box 41080
Los Angeles 90041-0080
(818) 567-1496

History/description: The hub of horticultural interest in Southern California since 1935, the Society offers something of value for every level of plant devotee. University professors and amateur gardeners share information at monthly meetings (second Thursday evenings) which include a speaker, plant raffle, plant forum, plant sale and occasional book sale.

SCHOOLS

The following California schools and colleges offer horticulture, landscape design and nursery managemnt courses. Most include practical training.

American River College
Sacramento
Dept.: Ornamental Horticulture
Degree: Certificate, AA

Antelope Valley College
Lancaster
Dept.: Ornamental Horticulture
Degree: Certificate, AA

Bakersfield College
Bakersfield
Dept.: Ornamental Horticulture
Degree: Certificate, AA

Butte College
Oroville
Dept.: Ornamental Horticulture
Degree: Certificate, AA

Cabrillo College
Aptos
Dept.: Greenhouse Management, Landscape Maintenance, Ornamental Horticulture
Degree: Certificate (GM, LM), AA (GM, OH)

California Polytechnic State University
Pomona
Dept.: Ornamental Horticulture, Landscape Architecture
Degree: BS (OH), BS, MLA (LA)

California Polytechnic State University
San Luis Obispo
Dept.: Ornamental Horticulture, Landscape Architecture
Degree: BS

California State Univeristy
Chico
Dept.: Ornamental Horticulture

California State University
Fresno
Dept.: Ornamental Horticulture
Degree: Concentration in another major

Cerritos College
Norwalk
Dept.: Landscape Maintenance, Ornamental Horticulture
Degree: Certificate (LM, OH), AA (OH)

City College of San Francisco
San Francisco
Dept.: Ornamental Horticulture, Retail Nursery
Degree: Certificate (RN), AA (OH)

College of Marin
Kentfield
Dept.: Landscape Construction, Landscape Maintenance, Nursery Management
Degree: Certificate, AA

College of San Mateo
San Mateo
Dept.: Ornamental Horticulture
Degree: Certificate

College of the Desert
Palm Desert
Dept.: Ornamental Horticulture
Degree: Certificate, AA

College of the Sequoias
Visalia
Dept.: Landscape Management, Nursery Management, Ornamental Horticulture
Degree: Certificate (all), AA (OH)

Cosumnes River College
Sacramento
Dept.: Landscape Design, Landscape Maintenance, Nursery Operations
Degree: Certificate

Covelo Farm School
Covelo
Dept.: Education in organic farming using primary agricultural methods, such as draft horses.
Degree: Two year apprenticeship

Cuyamaca College
El Cajon
Dept.: Landscape Tech, Nursery Tech
Degree: Certificate, AA

Diablo Valley College
Pleasant Hill
Dept.: Ornamental Horticulture
Degree: Certificate

El Camino College
Via Torrance
Dept.: Ornamental Horticulture
Degree: Certificate, AA

Foothill College
Los Altos Hills
Dept.: Landscape Horticulture, Nursery Management
Degree: Certificate

Fullerton College
Fullerton
Dept.: Greenhouse/Nursery Production, Landscape Design & Management
Degree: Certificate

Kings River Community College
Reedley
Dept.: Ornamental/Landscape Horticulture
Degree: Certificate, AA

Las Positas College
Livermore
Dept.: Ornamental Horticulture
Degree: Certificate, AA

Long Beach City College
Long Beach
Dept.: Landscape Construction, Landscape Design, Landscape Maintenance, Ornamental Horticulture
Degree: Certificate (all), AA (OH)

Los Angeles Pierce College
Woodland Hills
Dept.: Ornamental Horticulture
Degree: Certificate, AA

Mendocino College
Ukiah
Dept.: Horticulture, Landscape Practices
Degree: Certificate (LP), AA (H)

Merced College
Merced
Dept.: Landscape Horticulture
Degree: Certificate, AA

Merritt College
Oakland
Dept.: Landscape Horticulture, Landscape Construction & Design, Landscape Gardening & Maintenance, Nursery Management
Degree: Certificate (LCD, LGM, NM), AA (LH)

Mira Costa College
Oceanside
Dept.: Horticulture, Landscape Horticulture, Landscape Maintenance, Nursery Production
Degree: Certificate (LH, LM, NP), AA (H)

Modesto Junior College
Modesto
Dept.: Ornamental Horticulture
Degree: AA

Monterey Peninsula College
Monterey
Dept.: Ornamental Horticulture
Degree: Certificate

Mt. San Antonio College
Walnut
Dept.: Ornamental Horticulture
Degree: Certificate, AA

Schools

Napa College
Napa
Dept.: Ornamental Horticulture
Degree: Certificate, AA

Ohlone College
Fremont
Dept.: Landscape Horticulture
Degree: Certificate, AA

Orange Coast College
Costa Mesa
Dept.: Ornamental Horticulture
Degree: Certificate, AA

Saddleback College
Mission Viejo
Dept.: Landscape Design, Ornamental Horticulture
Degree: Certificate, AA

San Diego Mesa College
San Diego
Dept.: Nursery Architecture Technology, Nursery Landscape Technician
Degree: Certificate, AA

San Joaquin Delta College
Stockton
Dept.: Ornamental Horticulture
Degree: Certificate, AA

Santa Barbara City College
Santa Barbara
Dept.: Landscape Design, Nursery/Greenhouse Technology, Ornamental Horticulture
Degree: AA

Santa Rosa Junior College
Santa Rosa
Dept.: Landscape Management, Nursery Production
Degree: Certificate, AA

Shasta College
Redding
Dept.: Ornamental Horticulture
Degree: Certificate, AA

Sierra College
Rocklin
Dept.: Ornamental Horticulture
Degree: Certificate, AA

Solano Community College
Suisun City
Dept.: Ornamental Horticulture
Degree: Certificate, AA

Southwestern College
Chula Vista
Dept.: Landscape Architecture, Nursery Occupations
Degree: Certificate (NO), AA (LA)

University of California
Berkeley
Dept.: Landscape Architecture
Degree: BS, MLA

University of California
Davis
Dept.: Landscape Architecture, Horticulture
Degree: BS (LA), MS (H)

University of California
Santa Cruz 95064
Dept.: Agroecology Program - six months intensive hands-on study of ecological horticulture.
Degree: Certificate

University of California, Extension
Berkeley
Dept.: Garden Design, Landscape Architecture
Degree: Certificate

University of California, Extension
Irvine
Dept.: Landscape Architecture, Ornamental Horticulture
Degree: Certificate

Ventura College
Ventura
Dept.: Ornamental Horticulture
Degree: Certificate, AA

Victor Valley College
Victorville
Dept.: Ornamental Horticulture
Degree: AA

Yuba College
Marysville Campus, Woodland Campus
Dept.: Ornamental Horticulture
Degree: Certificate

■274 Where on Earth!

Master Gardener Programs

These University of California Cooperative Extension offices offer Master Gardener Programs, as well as good local information about what to grow and how to grow it.

Alameda County
224 West Winton Avenue
Room 174
Hayward, CA 94544
(510) 670-5200
FAX: (510) 670-5231

Amador County
108 Court Street
Jackson, CA 95642
(209) 223-6482
FAX: (209) 223-3312

Calaveras County
P. O. Box 837
San Andreas, CA 95299
(209) 754-6477
FAX: (209) 754-6472

Contra Costa County
1700 Oak Park Boulevard
Building A-2
Pleasant Hill, CA 94523
(510) 646-6540
FAX: (510) 646-6708

El Dorado County
311 Fair Lane
Placerville, CA 95667
(916) 621-5502
FAX: (916) 642-0803

Fresno County
1720 South Maple Avenue
Fresno, CA 93702
(209) 488-3285
FAX: (209) 488-1975

Kern County
1031 South Mt. Vernon Avenue
Bakersfield, CA 93307
(805) 861-2631
FAX: (805) 834-9359

Los Angeles County
2615 South Grand Avenue
Suite 400
Los Angeles, CA 90007
(213) 744-4851
FAX: (213) 745-7513

Marin County
1682 Novato Boulevard
Suite 150-B
Novato, CA 94947
(415) 899-8620
FAX: (415) 899-8619

Nevada County
Veterans Memorial Building
225 South Auburn Street
Grass Valley, CA 95945
(916) 273-4563
FAX: (916) 273-4769

Placer County
11477 E Avenue
Auburn, CA 95603
(800) 488-4308
FAX: (916) 889-7397

Riverside County
21150 Box Springs Road
Moreno Valley, CA 92387
(714) 683-6491
FAX: (714) 788-2615

Sacramento County
4145 Branch Center Road
Sacramento, CA 95827
(916) 366-2013
FAX: (916) 366-4133

San Bernardino County
777 East Rialto Avenue
San Bernardino, CA 92415
(714) 387-2171
FAX: (714) 387-3306

Master Gardener Programs

San Diego County
5555 Overland Avenue
Building 4
San Diego, CA 92123
(619) 694-2845
FAX: (619) 694-2849

Santa Clara County
2175 The Alameda
Suite 200
San Jose, CA 95126
(408) 299-2635
FAX: (408) 246-7016

Solano County
2000 West Texas Street
Fairfield, CA 94533
(707) 421-6790
FAX: (707) 429-5532

Sonoma County
2604 Ventura Avenue
Room 100-P
Santa Rosa, CA 95403
(707) 527-2621
FAX: (707) 527-2623

Stanislaus County
733 County Center III Court
Modesto, CA 95355
(209) 525-6654
FAX: (209) 525-4619

Sutter-Yuba Counties
142-A Garden Highway
Yuba City, CA 96991
(916) 741-7515
FAX: (916) 673-5368

Tuolumne County
2 South Green Street (mailing)
Sonora, CA 95370
(209) 533-5695
FAX: (209) 532-8978

Yolo County
70 Cottonwood Street
Woodland, CA 95695
(916) 666-8143
FAX: (916) 666-8736

Plant Index

Abies 18
Abutilons 8
Acacia 129, 210, 218, 231
Acer 14
    palmatum 48
Achillea 137, 155
Aechmea 108
African violet 110, 112, 125
Agave 187, 192, 237, 241
Alfalfa 263
Alliums 263
Aloe 44, 187, 192, 237, 241, 249
Alpine plants 4, 50, 79, 97, 137, 155, 159, 173, 255, 257
Alstroemeria 120, 217, 226
Amaryllis 201
Anemones 8
Anigozanthus 217
Annuals 24, 126, 154, 173, 177
Annuals seed 264
Apples 61
    Antique 74
Apples trees 7
Aquatic plants 49, 82, 104, 151, 173, 194, 235, 242
Arbutus 80
Arctostaphylos 3, 75, 160, 172, 180, 197, 239
Artemisia 126, 137, 155, 169, 243
Ash, Oregon 59
Asian medicinal plants 221
Aster 115
Astilbe 258
Astrophytum 199
Atriplex 231
Australian plants 47, 99, 109, 168, 172, 181, 189, 191, 192, 210, 220, 226, 228, 241
Azadirachta 86
Azaleas 17, 27, 55, 62, 81, 129, 145, 166, 168, 206, 230, 236, 257
Azara 192

Bamboo 33, 170, 221, 241
Bay 59
Beaucarnea 44
Bee plants 256
Beech 140
Begonia 5, 165, 168, 177
Berries 138
Bog plants 49
Bonsai 4, 46, 50, 76, 79, 112, 143, 146, 211, 235
Brachychiton 229
Brodiaea 263
Bromeliads 87, 108, 233
Buddleia 120
Bulbs 26, 78, 97, 105, 173, 208, 235, 261, 263
    Native 78
    South African 177
    South African 249, 261

Cacti 46, 56, 63, 100, 111, 122, 164, 171, 181, 192, 195, 199, 205, 219, 224, 232, 237, 241, 245, 249, 253
Cactus 203
Calluna 3
Calochortus 78, 263
Calodendrum 218
Camellia 8, 47, 80, 94, 110, 206, 213, 230, 235, 246
Campanula 13, 97, 115, 155, 159, 169, 173
Canna 89, 170, 208, 236
Carnivorous plants 35
Carpenteria 80, 27
Castanospermum 218
Casuarina 192
Cattleya 106, 107, 135, 193, 196
Caudiciforms 229
Ceanothus 3, 75, 80, 160, 172, 200, 239, 243, 248
Cercocarpus 160
Chamaecyparis 8
Cherry 92, 221
Chionanthus 27
Chorisia 229
Chrysanthemum 110, 209
Chrysothamnus 160
Cineraria 165
Cistus 2, 13, 39, 172, 210
Citrus 101, 253
Clematis 168
Clerodendrum 165
Clover 263
Coffea 192
Conifers 18, 53, 65, 69, 93, 160, 182, 245, 248, 255, 256
Conophytum 205
Copiapoa 195
Corn 263
Cornus 27, 145
Corylopsis 145
Corynocarpus 218
Corynodendron 218
Cover crops 259, 262
Crape myrtle 114, 142
Crassula 205, 241
Crinum 208
Cryptomeria 8
Cussonia 218
Cycads 44, 53, 203, 233, 247
Cyclamen 168, 200
Cyclamen seed 261
Cymbidium 135, 186, 193, 196

Daffodils 110, 123, 201
Dahlia 110, 112, 178
Daphne 2, 166
Daylilies 31, 34, 123, 216, 217, 223, 234, 255
Deer resistant plants 4, 137, 147, 155
Dendrobium 87, 193
Desert plants 152, 246, 247, 248
Desfontainea 27

Plant Index

Dianthus 115, 159, 173
Disanthus 27
Dogwood 80, 124, 140, 166
Drimys winteri 27
Drought tolerant plants 2, 4, 24, 39, 51, 59, 60, 75, 99, 110, 111, 116, 127, 129, 133, 137, 142, 144, 145, 147, 152, 155, 168, 170, 172, 183, 184, 191, 192, 203, 217, 220, 228, 231, 243, 249, 251
Dudleya 122, 192
Dwarf Conifer 8, 34, 50, 124, 206, 257
Dye plants 256

Echeveria 171, 187
Edible flowers 262
Embothrium coccineum 27
Enkianthus 'Red Bells' 27
Epiphyllum 216, 232
Erica 3
Erigonium 97
Eriocarpus 122
Erodiums 43
Erythrina 218
Eucalyptus 129, 182, 210, 256
Eucryphia 27
Euphorbia 44, 115, 171, 199, 205, 224, 237
Everlastings 256
Exochorda 27

Farm 71, 94, 277
Ferns 17, 90, 104, 165, 187, 189, 218, 244, 257, 258
Ficus 229, 249
Figs 221
Fire resistant plants 155, 231
Flower seed 259, 261, 262, 264
Flowers 71
Forb seed 260
Fothergilla 27
Fouquieria 122
Franklinia 27
Freesia 120
Fremontodendron 80
Fruit trees 7, 65, 114, 117, 138, 147
    Antique 61
    Espalier 71, 74, 117
    Subtropical 245
    Tropical 221, 241
Fruiting plants 256
Fuchsias 6, 19, 168, 181, 200, 249

Galvesia 231
Gardenias 208
Garrya 27
Geradanthus 237
Geraniaceae 64
Geraniums 8, 13, 39, 43, 115, 155, 159, 168, 170
Gingers 208, 238
Gingko 140, 166
Grains 259
Grapes 138
Grass seed 259, 260, 262, 263

Grasses 3, 8, 9, 34, 45, 52, 53, 70, 75, 100, 102, 104, 144, 170, 173, 206, 222, 258
    Native 184
Grasslike plants 53
Grevillea 189, 210
Ground Covers 9, 13, 59, 75, 114, 141, 169, 231
Guavas 221, 241
Gymnocalycium 171, 199

Halesia 27
Hamemelis 145
Hardy plants 39, 60, 69, 137, 139, 140, 142, 152, 154, 155, 159, 161, 162, 183
Haworthia 44, 171, 195, 205
Heat tolerant plants 139, 140, 154, 243
Heaths and Heathers 3, 8, 9, 17, 28
Hebe 39
Helianthemum 2
Helleborus 13, 206
Herb seed 259, 260, 261, 262, 264
Herbs 2, 29, 39, 66, 70, 71, 77, 86, 139, 147, 170, 225, 228, 256
Hibiscus 238, 247
Hostas 79, 255, 258
Houseplants 105, 126, 181, 219
Hoya 207, 232
Hymenocallis 208

Iris 7, 83, 89, 92, 102, 110, 112, 118, 121, 123, 128, 134, 208, 215, 217, 234, 244, 255
Ivies 28

Japanese Maples 8, 14, 51, 69, 114, 124, 140, 143, 206
Jasmine 238
Jujube 221

Kalmia 27

Laelia 106
Lagerstroemia 142
Larix 18
Lavandula 2, 8, 13, 126, 137, 172, 173, 225
Legume seed 259, 260
Leucodendron 75
Leucothoe 27
Lewisia 122
Lilies 5, 82, 112
Longani 221
Lotus 194
Luma 192
Lycaste 193
Lychee 221
Lycoris 208

Madagascar natives 44
Madrone 59, 80
Magnolia 105, 140, 166, 206, 256
Mahonia 239
Mammillaria 171, 195, 199
Mandevilla 165

Manzanita 80, 248
Maples 27, 48, 80, 146, 255
Markhamia 218
Masdevallia 106, 186
Mediterranean plants 2, 4, 9, 13, 36,
    58, 60, 94, 100, 109, 137, 172,
    181, 189, 191, 192, 197, 220, 221,
    226, 228, 243, 249, 251
Melaleuca 192, 218
Michelia 27, 235
Miltonia 196
Mimulus 210
Mint 225
Mulberries 221
Mushrooms 260

Nandina 206
Narcissus 15, 201
Native plants 24, 32, 34, 36, 37, 45,
    59, 60, 69, 70, 75, 80, 90, 93,
    102, 103, 110, 111, 116, 127, 129,
    133, 142, 144, 155, 161, 162, 168,
    170, 172, 174, 180,181, 184, 189,
    191, 192, 197, 200, 202, 210, 228,
    239, 243, 244, 247, 256, 263, 276
  Channel Islands 253
  Coastal 52, 198, 218, 240, 248
  Foothill 141, 151, 153
  High elevation 152, 160
  Montane 158
  North coast 2, 16, 24
  Riparian 248
  S.F. Bay Region 84
  Valley 151, 153
  Woodland 9, 198, 248
Neem 86
Neochilenia 195
Neoporteria 195
Neoregelia 108, 233
Notocactus 171
Nut trees 117, 138, 147
Nymphaea 194

Oaks 22, 37, 80, 88, 129, 133, 136,
    142, 213, 244
Odontoglossum 185
Oncidium 87, 106
Onions 262
Opuntia 199
Orchids 19, 26, 57, 85, 87, 105, 106,
    107, 112, 135, 168, 177, 185, 186,
    193, 196, 261

Pachycormis 237
Pachypodium 44
Palm seed 259
Palms 53, 65, 192, 199, 208, 218, 221,
    241, 245, 247, 249, 253
Paphiopedilum 19, 87, 185, 186, 196
Paphiophyllum 107
Parodia 199
Peaches 61
Pears 61
  Antique 74
Pedilanthus 237

Pelargoniums 6, 11, 64, 175, 225
Penstemon 13, 39, 97, 104, 133, 137,
    155, 169, 210, 226
Peony 255
Peppers 264
Perennials 3, 4, 8, 9, 13, 24, 28, 34,
    36, 38, 39, 47, 49, 52, 58, 68, 69,
    75, 79, 103, 104, 105, 114, 115,
    120, 126, 137, 138, 139, 142, 144,
    145, 147, 151, 152, 154, 155, 161,
    162, 168, 169, 170, 173, 177, 178,
    183, 184, 187, 191, 192, 197, 200,
    208, 217, 220, 226, 228, 241, 243,
    255, 256, 258
Perennials seed 264
Phalaenopsis
    87, 107, 135, 185, 193, 196
Phlox 126, 159, 173
Picea 18
Pieris 166
Pines 8, 28, 146, 152, 249, 252
Pleione 186
Plumcot 72
Plum 61
Poinsettia 200
Pomegranate 221
Primula 97
Protea 75, 189, 203
Prunus 160
Purshia 160
Pygmy forest 29

Quercus 22
Quince 4

Rare plants 26
Rebutia 171, 195
Redbud 75, 136, 147
Redwoods 29
Rhipsalis 232
Rhododendrons 17, 20, 21, 23, 25, 27,
    28, 29, 40, 41, 55, 62, 81, 166,
    176, 178, 256, 257
Riparian plants 152, 158
Rock garden plants 4, 9, 15, 79,
    104, 110, 131, 137, 257, 258
Rosemary 2, 13, 225
Roses 7, 42, 67, 71, 91, 92, 95, 98, 104,
    105, 112, 130, 147, 148, 188, 213,
    241, 246, 249, 252
  Miniature 50, 150, 212
  Old 7, 10, 12, 67, 98, 149,
    169, 188, 190, 212

Salt tolerant plants 231
Salvia 13, 86, 104, 115, 169, 170, 173,
    183, 226, 241
Sansevieria 224, 229
Santolina 137
Schlumbergia 232
Sedum 63, 122, 159, 237
Sempervivum 63, 122
Shade plants 59, 165, 258
Shade trees 147
Shiitake mushrooms 260

Plant Index

Shrub seed 259, 260, 261, 263
Shrubs 3, 4, 32, 34, 39, 62, 65, 68, 75, 77, 80, 98, 102, 124, 138, 140, 141, 143, 144, 145, 151, 154, 161, 166, 181, 197, 218, 231, 235, 241, 256, 257, 258
  High elevation 160
  Native 182
  Subtropical 98
South African plants 47, 93, 99, 172, 189, 191, 210, 220, 241
Spathodea 218
Staghorn Ferns 233
Stenocarpus 218
Stewartia 27
Styrax 27
Subtropical plants 54, 99, 217, 220, 221, 238, 251
Succulents 44, 46, 56, 63, 93, 100, 111, 122, 164, 170, 171, 173, 187, 192, 195, 199, 203, 205, 219, 224, 229, 237, 241, 245, 253

Thunbergia 238
Thyme 137, 173, 225
Tillandsia 108, 233

Toyon 80, 136
Tree seed 259, 260, 261, 263
Trees 3, 4, 13, 32, 34, 52, 62, 65, 68, 69, 75, 80, 88, 124, 127, 129, 130, 138, 140, 143, 144, 145, 151, 154, 161, 166, 172, 174, 181, 182, 197, 213, 218, 229, 231, 241, 255, 256, 257
  High elevation 160
Tropical plants 174

Vegetable seed 256, 259, 261, 262, 263, 264
Vegetables 24, 29, 71, 154
Verbena 126
Viburnum 27
Vines 13, 68, 238
Violet, sweet 115
Vitex 218
Vriesea 108

Water Iris 49
Water Lilies 49
Wildflower seed 259, 260, 262, 263
Wisteria 79, 92, 94, 253

Zygopetalum 106, 186

Growers

Abbey Garden Cacti & Succulents ...164
Albright Seed Company 259
Alpine Valley Gardens 31
Alpines 97
Anderson Valley Nursery 2
Anhorn Nursery 103
Antonelli Brothers Begonia Gardens .165
Appleton Forestry 32
Arbor & Espalier Co. 74

Baccus, C. H. 78
Baltzer's Specialty Plants 255
Bamboo Sourcery 33
Barfod's Hardy Ferns 257
Bay Laurel Nursery 166
Baylands Nursery 75
Bell Hardware & Nursery 114
Berkeley Horticultural Nursery, Inc. ...98
Berkeley Municipal Rose Garden 112
Bidwell Park 130
Bio-quest International 167
Blake Garden 109
Blooming Fields Farm 215
Blooming Nursery 255
Blossom Valley Gardens 216
Blue Oak Nursery 133
Bluebird Haven Iris Gardens 134
Bonsai Garden 76
Borees Nursery 257
Bountiful Gardens 259
Bourne Mansion 156
Briarwood Nursery 77
Brinsley's Orchids 135
Buena Creek Gardens 217
Burgandy Hill Nursery 34

Cactus By Dodie 205
Calaveras Nursery 136
California Carnivores 35
California Flora Nursery 36
Canyon Creek Nursery 115
Capitol Park 129
Caprice Farms Nursery 255
Carman's Nursery 79
Carmel Valley Begonia Gardens 168
Carter Seeds 259
Central Coast Growers 169
Chavez Ravine Arboretum 250
Chia Nursery 170
Choice Edibles 260
Christensen Nursery Company 80
Circuit Rider Productions, Inc. 37
Clyde Robbin Seed Company 260
Coastal Zone Nursery 218
Collector's Nursery 255
Conservaseed 260
Copacabana Gardens 99
Cornflower Farms 116
Costa Nursery 117
Cottage Garden Growers 38
Cottage Gardens 118
Covered Bridge Gardens 119

Davis Arboretum 129
Descanso Gardens 244
Desert Images 219
Desert Theatre 171
Desert to Jungle Nursery 220
Digging Dog Nursery 3
Drought Resistant Wholesale Nursery 172
The Dry Garden 100
Dunsmuir House 112
Dynasty Gardens 173
"El Molino Viejo," Old Mill 251
Emerisa Nursery 39
Enjoy Rhododendrons 40
Environmental Seed Producers 260
Evergreen Gardenworks 4
Ewing Orchids 179
Exotica Rare Fruit Nursery & Seed
  Company 221

Fairyland Begonia & Lily Garden 5
Fancy Fronds 258
Farwell & Sons Nursery 41
Farwell's Nursery 81
Farwest Nurseries 174
Fetzer Vineyards Garden Project
  At Valley Oaks 29
Filoli Center 91
Flowers & Greens 120
Foothill Cottage Gardens 137
Forestfarm 256
Forest Seeds of California 261
Four Winds Growers 101
Fowler Nurseries, Inc. 138
Fox Run Nursery 139
Franklin Murphy Sculpture Garden .251
Fuchsiarama 6
Fullerton Arboretum 245

Gamble Garden Center 92
Gamble House 250
Garden Valley Ranch Nursery 42
Genetic Resource Center 130
Geraniaceae 43
Getty Museum Gardens 250
Gold Hill Nursery 140
Goethe Arboretum 130
Gold Run Iris Garden 121
Golden Gate Park 93
Goodwin Creek's Secret Garden 256
Gossler Farms Nursery 256
Grand Ridge Nursery & Pottery 258
Grasslands Nursery 102
Geranium House 175
The Great Petaluma Desert 44
Green Gulch Farm 71
Green Leaf In Drought Time 141
Greenlady Gardens/Skittone Bulb Co. 261
Greenlee Nursery 222
Greenmantle Nursery 7
Greenwood Daylilies 223
Greer Garden 256
Grigsby Cactus Gardens 224

Hakone Japanese Gardens 94
Hannah Carter Japanese Garden 250

Grower Index

Haver's Rhodendrons 176
Heartwood Nursery 8
Heather Farm Garden Center 109
Heirloom Garden Seeds 261
Henderson's Experimental Gardens ..206
Herban Garden Herbs 225
Heritage House Nursery 9
Heritage Rose Garden 10
Heronswood Nursery 258
Hi-mark Nursery, Inc. 177
High Ranch Nursery 142
Hill 'n Dale 207
Hope Rhododendron Nursery 178
Hudson, J. L., Seedsman 261
Huntington Library & Botanical Gardens
.............................245

Independence Trail 156
Instant Oasis 82

Jensen Botanic Garden 130
Judy's Perennials 226
Jughandle State Reserve 29

Kartuz Greenhouses 227
Kelly's Plant World 208
King's Mums 209
Kitazawa Seed Company 261
Korbel Champagne Cellars Garden 72
Kruse Rhododendron State Reserve ...29

La Casita De Arroyo 250
Lake's Nursery 143
Lakeside Park Center & Gardens 112
Larner Seeds 45
Las Pilitas Nursery 180
Leaning Pine Arboretum 203
Rowntree, Lester, Arboretum 203
Lockhart Seeds 262
Lone Pine Gardens 46
Longden Nursery 47
Lorraine's Geranium Collection & Woods
Farm Nursery 11
Los Angeles State & County Arboretum
.............................246
Los Osos Valley Nursery 181
Lotus Valley Nursery & Gardens 144
Lotusland 203
Lummis House, "El Alisal" 251
Luther Burbank Home & Gardens 71
Lyon Tree Farm 182

Magic Gardens 104
Magic Growers 228
Maple Leaf Nursery 145
Marca Dickie Nursery 48
Marin Art & Garden Center 72
Markham Nature Park & Arboretum ..112
Maryott's Iris Garden 83
Matsuda Landscape & Nursery 146
Maxim's Greenwood Gardens 123
McAllister Water Gardens 49
Meadowlark Nursery 183
Mendocino Coast Botanical Gardens ..28
Mendocino Heirloom Roses 12
Mendon's Nursery 124
Menzies Native Nursery 158

Merritt College, Ornamental Horticulture
Department 105
Micke Grove Park 213
Mighty Minis 125
Mildred E. Mathias Botanical Garden .251
Miniature Plant Kingdom 50
Momiji Nursery 51
Montgomery Woods State Reserve 29
Moon River Nursery 13
Morcum Amphitheatre of Roses 112
Mostly Natives Nursery 51
Mountain Maples 14
Mt. Tahoma Nursery 258
Muchas Grasses 53

Native Revival Nursery 84
Native Sons Wholesale Nursery 184
Native Uprisings 16
Neon Palm Nursery 53
Nichols Nursery 256
North Coast Rhododendron Nursery ...55
Northridge Gardens 229
Noyo Hill Nursery 17
Nuccio's Nurseries 230

Oasis 56
Old Mill, "El Molino Viejo" 251
The Orchid House 185
The Orchid Ranch 107
Orchidanica 106
Orchids of Los Osos 186
Ornamental Edibles 262
Overfelt Botanical Gardens 94

Pacific Coast Seed Company 262
Peaceful Valley Farm Supply 262
Pecoff Bros. Nursery & Seed 231
Petite Plaisance - Orchids 57
Pianta Bella 187
Planet Earth Growers Nursery 210
The Plant Barn 126
Plants from the Past 58
Pomona College 252
Porterhowes Farms 257
Prusch Farm Park 94

Quail Botanical Gardens 247

Rainbow Gardens Nursery
  & Bookshop 232
Rainforest Flora 233
Ramsey Seeds 263
Rancho Los Alamitos 252
Rancho Santa Ana Botanic Garden ...248
Rare Conifer Nursery 18
Red's Alpine Gardens 257
Redbud Farms 147
Redwood City Seed Company 263
Regine's Fuchsia Gardens & The Orchid
  Bench 19
Regional Parks Botanic Garden 110
Stretz, Richard, Hybridizer 234
Richards Landscape & Gardening 20
River Oaks Nursery 127
Robinett Bulb Farm 263
Robyn's Nest Nursery 258
Mclellan, Rod, Co., Acres of Orchids ..85

Roeding Park 213
Roris Gardens 128
Rose Acres 148
Rose Garden in McKinley Park 130
The Rose Ranch 188
Rosendale Nursery 189
Roses of Yesterday & Today 190
Roses & Wine 149
Russ Quality Plants 122
The Ruth Bancroft Garden 110

S. F. League of Urban Gardeners 92
S & S Seeds 263
San Gabriel Nursery 235
San Jose Municipal Rose Garden 95
San Marcos Growers
   Wholesale Nursery 191
San Mateo Garden Center 94
San Mateo Japanese Tea Garden 95
San Miguel Greenhouses 108
San Simeon Nursery 192
Santa Barbara Botanic Garden 202
Santa Barbara Orchid Estate 193
Santa Barbara Water Gardens 194
Sarver Nursery 236
Saso Herb Gardens 86
Scott Valley Perennials 159
Seiju-en Bonsai Nursery 211
Sequoia Nursery 212
Serra Gardens 237
Shein's Cactus 195
Shelldance Nursery 87
Shepherd Garden & Art Center 131
Shepherd's Garden Seeds 263
Sherman Library & Gardens 252
Sherwood Nursery 21
Shrub Growers Nursery 59
Sierra Foothill Miniature Roses ..150
Sierra Valley Nursery 160
Siskiyou Rare Plants 257
Skylark Wholesale Nursery 60
Sonoma Antique Apple Nursery 61
Sonoma Horticultural Nursery 62
South Coast Botanic Garden 249
Specialty Oaks 22
Spotted Flower Nursery 151
Stallings Ranch Nursery 238
Stewart Orchids 196
Sticky Business 63

Strybing Arboretum
   & Botanical Gardens 93
Summers Lane Nursery 23
Suncrest Nursery 197
Sunset Coast Nursery 198
Sweetland Farm Wholesale Nursery ..152

Tahoe Tree Company 161
Taylor's Greenwood Nursery 199
Territorial Seed Company 264
Theodore Payne Foundation 239
Tiedemann Nursery 200
Tierra Madre 24
Torrey Pines State Reserve 252
Tree of Life Nursery 240
Trees of California 88
Trillium Lane 25
Tropic World 241

U. C. Botanical Garden 111
U.C. Irvine Arboretum 249
U.C. Riverside Botanical Garden ..252
U.C. Santa Cruz Arboretum 202
Up Sprout Geraniums 64
Urban Tree Farm 65
Urmini Herb Farm 66

Van Ness Water Gardens 242
Villa Montalvo 95
Villager Nursery 162
Vintage Gardens 67
Virginia Robinson Gardens 252

Watchwood Garden Design
   & Nursery 26
Weiss Brothers Nursery 154
Welch, William R. P. 201
Western Hills Nursery 68
Westgate Garden Nursery 27
Wildwood Farm 69
Wildwood Nursery 243
William Land Park 131
Wilson, Nancy R., Species & Miniature
   Narcissus 15
Wisteria Vine 253
Wright Iris Nursery 89
Wrigley Memorial Botanic Garden ..253

Ya-Ka-Ama Nursery 70
Yerba Buena Nursery 90
You Bet Farms 155